BIOSPHERE
The Human 2 Experiment

John Allen
Edited by Anthony Blake

PENGUIN BOOKS

PENGUIN BOOKS
Published by the Penguin Group
Viking Penguin, a division of Penguin Books USA Inc.,
375 Hudson Street, New York, New York 10014, U.S.A.
Penguin Books Ltd, 27 Wrights Lane,
London W8 5TZ, England
Penguin Books Australia Ltd, Ringwood,
Victoria, Australia
Penguin Books Canada Ltd, 2801 John Street,
Markham, Ontario, Canada L3R 1B4
Penguin Books (N. Z.) Ltd, 182–190 Wairau Road,
Auckland 10, New Zealand

Penguin Books Ltd, Registered Offices:
Harmondsworth, Middlesex, England

First published in simultaneous hardcover and paperback editions
by Viking Penguin, a division of Penguin Books USA Inc. 1991

1 3 5 7 9 10 8 6 4 2

A SYNERGETIC PRESS, INC., PRODUCTION

Written by John Allen with:

Kathleen Dyhr
Mark Nelson
Linnea Gentry
Tom Dollar
Mark Turner

Anthony G. Blake, *Editor-in-Chief*
Deborah Parrish Snyder, *Executive Editor*
David Stanford, *Viking Penguin Editor*

ISBN 0 14 01. 5392 6
CIP data available

Graphic design and production by Debra Kay Niwa
Photographic research by M.M. Evans
Cover photos: *top:* © 1991 Peter Menzel;
inset left: Gill C. Kenny; *inset middle:* C. Allan Morgan; *inset right:* Jeff Topping
Printed in the United States of America by Arizona Lithographers, Tucson

Printed on recycled paper

Table of Contents

To
Lynn Margulis
Clair Folsome
and
Evgenii Shepelev

Introduction

The self-contained sealed environment known as *Biosphere 2* now stands glinting in the sun of the Sonoran desert, at the start of what scientists plan to be a hundred-year lifetime. The four men and four women inside it at the beginning of 1991 are committed to a two-year experiment in life and living. They, the support team around them and the numerous specialists on call, are determined to make it work. Of course, the future is unknown and in science it is almost certain that what actually happens will surprise us, and not only in pleasant ways. We have no way of predicting exactly which and how many of the thirty-eight hundred species Biosphere 2 contains will survive. All that we do know is that the technology will probably keep going and the people, hopefully, will not go crazy. Everything has been done to anticipate and compensate for all the breakdowns we can conceive as possible; but nothing can insure against the inconceivable. What Biosphere 2 will mean and reveal and prove in the end we can only guess at. Even if the entire experiment dissolves into green slime, as one of our consultants jokes, it will have been worth the effort involved because we will have measured the events and gained many new insights for the next steps. This book is the story of how Biosphere 2 was conceived, developed and built, and what we hope it will teach us.

The story of Biosphere 2 begins ten thousand years ago with the origins of agriculture, when man first began making human-controlled ecosystems — or 'farms' as they are better known — and exploited the surplus energy thus made available to create cities and develop technology. Man's impact on the natural world has increased since that

Biosphere 2: an aid to dealing with the problems of the environment; an experiment to understand the laws of biospherics; and a prototype for a space colony.

1

time at what seems to be an ever-accelerating rate.

It was not until the early part of this century that we developed a clear idea of what we now call the biosphere. Professor Vladimir I. Vernadsky used the term scientifically for the first time almost seventy years ago to describe the zone of life on planet Earth — the region extending several miles into the sky, and several miles below the Earth's crust. It is within this region that the biosphere, or "sphere of life" exists. As mankinds' actions have had more and more effect on the intricate web of systems that tranform energy and support life, and as our awareness of these systems increased, it has become clear that all of life on Earth is a vast coordinated system. As we observed more and more, it became clear that even the smallest elements — the microbes, the invisible hordes of miniscule creatures living in soil, water, the atmosphere — play key roles in managing the chemistry of life on Earth.

As human's effects on the system have proved deadly to many creatures, and, as we begin to realize, may prove deadly for humankind as well, attention has been turning towards means of integrating our technological advancements with the delicate web of life that supports us.

A group of individuals began in the late 60s and early 70s to work with the idea of developing a synergy between the "ecology of technics and the technics of ecology" and formed the Institute of Ecotechnics. The name 'ecotechnics' is like the name 'ecology' or 'economics', in that the 'eco' signifies the 'household', but in this case the household is that of the whole biosphere. For nearly two decades, the Institute of Ecotechnics' directors, members, and consultants implemented projects in areas of environmental stress or mis-management to test their capabilities of integrating the technosphere with the biosphere.

As this vision grew, the Institute began to study *biomes* — rainforests, savannahs, oceans, deserts — and perceived biomes to be the main ecological "building blocks" of the biosphere. The focus of the Institute's efforts then moved to developing total systems management programs to deal with biomic scale problems such as desertification in semi-arid tropical savannah regions, deforestation and over-cutting in the rainforest, the Mediterranean farm threatened by industrial and urban expansion, and pollution and overexploitation in the world ocean.

With the dawn of the space-age came the view of Earth from space that enabled many people to see immediately, instinctively, that there was only one Earth and that it was finite and precious. The prospect of travel to other worlds began to make us realize that the biosphere of Earth was unique in our solar system and we would go nowhere off this planet for long without a similar life support system.

It had only become possible in the last half of the twentieth century to have a sufficient concentration of wealth — resources and energy — as well as an adequate technology for a launch into space. At the same time, problems escalated here on the ground, especially as the population expanded at an explosive rate. It looked like there might be only a fairly narrow 'window of opportunity' before urgent terrestrial demands completely swamped any prospect of funds being allocated to travel to other worlds. Increasingly, it looked like a dichotomy: sort out things on Earth OR go into space and ignore the mess left behind. As with many dichotomies, the apparent choice was spurious.

The evolution of the theory of the biosphere combined with the fifteen years of experience working in the field by several of the Institute of Ecotechnics' directors helped to give rise to the decision to build a demonstration model of a biosphere. This later developed into the concept of a landmark enterprise — Space Biospheres Ventures — whose corporate objective would be to design, build, and operate closed ecological systems for biospheric research and education towards better management on Earth, for applications in the exploration and settlement of the solar system — and perhaps beyond.

We — and many other people, too — noticed that most of the thinking about space-travel dealt with the problems of supplying the bare necessities to support life, following guidelines based on such things as conditions in nuclear submarines. If we could work towards a viable self-contained environment for people to live a full life in on other worlds, then we would not only help make sense of space travel, but also learn things to help us here on Earth. The Russians had by the seventies made some important advances in this area. In America, the Odums, both penetrating total systems ecologists, had produced major advances in ecological theory. Work was already in hand. It was time to pull the pieces together and help make a better future.

In 1984 NASA plans called for Space Station *Freedom* to be in orbit in 1992. Space Biospheres Ventures drove to get Biosphere 2 built and into operation by that date, anticipating the possibility of putting the first small space life system into orbit by 1995. Venture capital was raised on the assumption that marketable technology would be developed, which would offer practical solutions to specific problems of pollution control and environmental management on Earth. The Biosphere 2 project would be not just a matter of science and technology, important as they were, but also one of appropriate finance, management, and product development.

A project of this kind depends absolutely on a core of people with sufficient insight and ability. Such people cannot be advertised for: they choose themselves and develop on the job. Biosphere 2 is a work of art as much as a work of engineering or a scientific experiment. The very nature of the Biosphere 2 project demands an integration of many disciplines: science, engineering, business, management, art, people, cybernetics, agriculture, east and west, north and south, new and old.

There is nothing radically new in any one of the individual things we have done. What has been extraordinary is to do all of these things all at once. Even with the best experts in the world contributing, we, and they, had to learn a lot by doing. What now may look like obvious application of theory or common sense has often been won through much sweat and tears. The full story of even a single successful step in the project would take too long to tell. What we have done is to give a cross section of the process of design, a sampling of both the grand cosmic ideas and the day to day things, the people as well as the theories, the problems as well as the breakthroughs. Remember that the experiment is 'in progress' and, even as you read, is revealing new problems and yielding new insights; so much so, that we may never catch up with what we have set in motion.

John Allen
November 22, 1990

1
The Pioneers

"At Delphi the priestess began her formal ritual address to the gods thus: 'First in my prayer before all other gods, I call on Earth, primeval prophetess'."

Charlene Spretnak

The Theory

The concept of the *biosphere* was born almost seventy years ago, mainly through the work of the Russian geochemist Vladimir Vernadsky (1863-1945). Vernadsky was a generalist whose work illuminated some of the interplay between geology, physics, chemistry and biology in the natural world. Versed in some fifteen languages, he had read the work of the Austrian geologist Eduard Suess who first introduced the term *biosphere* in 1875 to describe the 'envelope' around the planet which was inhabited by life. Vernadsky traced the seeds of this idea to the observations of the French naturalist Jean Baptiste Lamarck.

Vernadsky is a figure that looms large in the Soviet pantheon of scientists, held there in somewhat comparable esteem as Darwin is in European and American science. Some scientists consider that Vernadsky did for biological space what Darwin did for biological time. The work of both is necessary to understand biospherics. Darwin proved the unity of all life throughout the billions of years of time and the complexity of forms. Vernadsky showed the unity of all life in space, and that it operated on a daily scale as a cosmic phenomenon and geological force.

Vernadsky defined the biosphere as "the environment in which we live, it is the 'nature' that surrounds us and to which we refer in common parlance". He observed that as life evolves, it actually changes the environment in which it evolves. Increasingly complex forms of life appear, using larger amounts of matter and energy and converting them into new patterns of life. The ancient Greek philosopher Heraclitus noted that "being is ever becoming". To Vernadsky, dynamism, change, and flux were the linchpins of life.

In 1926, Vernadsky published *The Biosphere*, a volume in which he offered a compelling theoretical framework for the existence of a global ecosystem controlled by life. More than thirty years before humankind could obtain an extraterrestrial view of our planet, he wrote:

"The surface of the Earth, seen from the depths of infinite celestial space, seems to us unique, specific and distinct from that of all other heavenly bodies. The surface of our planet, its biosphere, separates the Earth from its cosmic surroundings. The terrestrial face becomes visible where it receives light from celestial bodies, particularly the Sun ... Under the influence of these light-rays, vegetable organisms produce chemical compounds which would be unstable in any other environment but that of the interior of the plant. The whole living world is connected with this green mass by a direct and indissoluble bond. We may, then, regard

MARIE ALLEN

Trees, mountains, clouds; winds, rains, gravity, surface tension; some of the dynamic forms and forces of the biosphere.

Opposite: Vernadsky was famous for his character and being an inspiration as well as for his deep and far-reaching thought.

living matter in its entirety as the peculiar and unique domain for the accumulation and transformation of the luminous energy of the Sun ... There is no force on the face of the Earth more powerful in its results than the totality of living organisms ... If life were to disappear, a stable chemical equilibrium, a chemical calm, would be established ... Without life, the face of the Earth would become as motionless and inert as the face of the Moon. Life, therefore, exerts a powerful permanent and continuous disturbing effect on the chemical stability of the surface of our planet. With its colors and forms, its combinations of vegetable and animal organisms and the creative activity of civilized humanity, life not only creates the whole picture of our natural surroundings but penetrates into the deepest and most grandiose processes in the Earth's crust."

According to Vernadsky, the evolution of human beings gave rise to two new dimensions in the biosphere: the *technosphere* and the *noosphere*. The technosphere was created with the evolution of technologies in agriculture, transportation, industry, genetics, and all the other areas of human activity. Vernadsky defined the noosphere as the region of thought and creativity where reasoning, remembering, and decision-making take place. The evolving noosphere of the human mind, he believed, would mediate and reconcile conflicts between the technosphere and the biosphere and help to direct further evolution, placing humans in the role of stewards of the Earth.

Unfortunately, politics, in the past, restricted communications between scientists in the Soviet Union and the West which could explain why relatively few Western scientists were familiar with Vernadsky's theory of the biosphere. Nor were his ideas heavily propagated within the Soviet Union while he was alive, as his emphasis on the power of life as a total system did not exactly scientifically support the political structure of his day. Nevertheless, his work served as the cornerstone for several large Soviet research institutions, founded and directed by Vernadsky, which continue to the present day.

Fortunately, Yale zoologist G. Evelyn Hutchinson encountered Vernadsky's work through meeting the Russian scientist's son, George Vernadsky, who emigrated to America and became a professor of history at Yale. In the mid 1940s, Hutchinson published a paper discussing the relevance of Vernadsky's work to biogeochemistry, but it went relatively unnoticed. Hutchinson integrated the insights of the biosphere hypothesis and moved ahead with his own groundbreaking research to become a pivotal figure in the development of ecological science. In 1969, *Scientific American* published the influential volume entitled *The Biosphere*, for which Hutchinson wrote the introduction. This publication introduced the name and work of Vernadsky to John Allen, Mark Nelson and their colleagues who would later conceive of the idea for Biosphere 2.

Theories of the evolution of life on Earth had previously cast life as the opportunistic but fragile passenger and chance beneficiary of benevolent circumstances on the planet. Vernadsky and his school offered a revolutionary theory — that it was *life itself* that made nearly all conditions necessary for its own evolution. Vernadsky hypothesized that all life on Earth is a single entity and that life itself manipulates the planetary environment by the transfer of matter on an awesome scale.

Take the example of the Earth's atmosphere. The envelope of air which surrounds our planet is about eighty percent nitrogen and twenty percent oxygen, with traces of methane, water vapor, and carbon dioxide. Scientists have long known that this mixture is chemically unbalanced. In theory, both nitrogen and oxygen should be quickly bound up in various chemical reactions with many different compounds on Earth; and methane is so chemically active it should not be present at all!

The great British atmospheric scientist, James Lovelock, made the next advance in biospheric theory by showing that Mars, Venus, and Jupiter were physical-chemical planets, oxidized to over ninety-five percent CO_2 on Mars and Venus, reduced to ammonia on Jupiter. Only the planet Earth kept an atmosphere with full oxidizing and reducing nitrogen and methane at the same time! Only Earth had life. Therefore the creation of an atmosphere particularly useful to life, with this tremendous amount of extra energy, was due to life itself working on the physical-chemical processes.

Evidence in the geologic record indicates that Earth's early atmosphere was quite different from what it is now, consisting almost entirely of carbon dioxide with only traces of nitrogen and no free oxygen. According to biospheric theory, the switch from a predominantly carbon dioxide atmosphere to an oxygen atmosphere is a dramatic example of how life on Earth has changed the environment. Ancient bacteria used hydrogen sulfide and released sulphur as waste. Other bacteria evolved that used water as their source of hydrogen and released *oxygen* as waste. They began to grow, powered with energy from the Sun, by combining hydrogen from water with carbon from the carbon dioxide in the air.

As these blue-green bacteria thrived, they removed large amounts of carbon dioxide from the atmosphere; at the same time oxygen released by these bacteria as waste also increased enormously. At first the free

Apollo 16 astronaut Charles M. Duke Jr. collects lunar rock samples April 1972.

Major cycles of the biosphere are indicated in a general way in this illustration. The operation of the biosphere depends on the utilization of solar energy for the photosynthetic reduction of carbon dioxide from the atmosphere to form organic compounds on the one hand and molecular oxygen on the other.

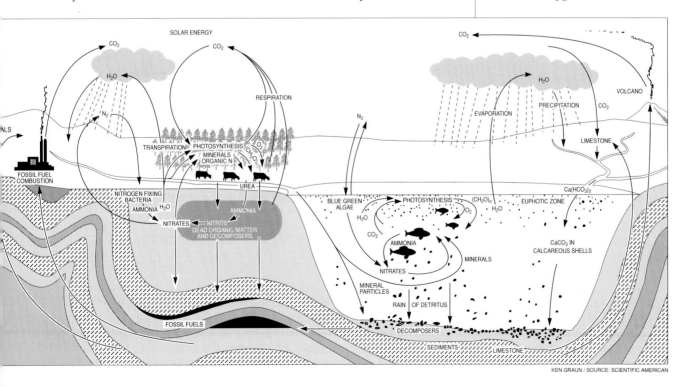

KEN GRAUN / SOURCE: SCIENTIFIC AMERICAN

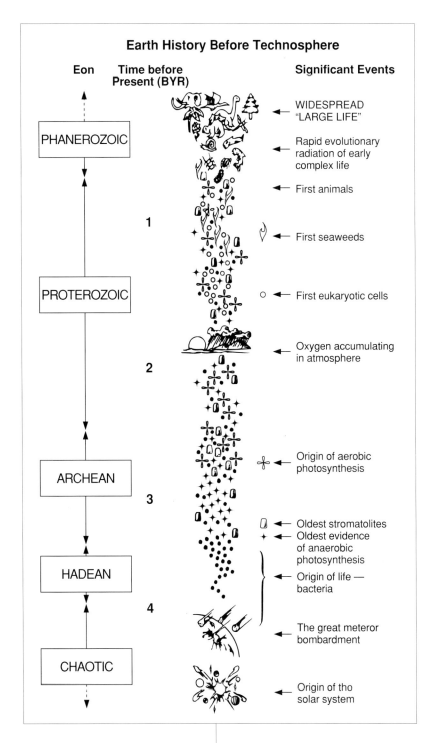

Earth History Before Technosphere

Eon	Time before Present (BYR)	Significant Events

PHANEROZOIC

← WIDESPREAD "LARGE LIFE"

← Rapid evolutionary radiation of early complex life

← First animals

1

← First seaweeds

PROTEROZOIC

← First eukaryotic cells

← Oxygen accumulating in atmosphere

2

ARCHEAN

3

← Origin of aerobic photosynthesis

← Oldest stromatolites
← Oldest evidence of anaerobic photosynthesis

HADEAN

← Origin of life — bacteria

4

← The great meteor bombardment

CHAOTIC

← Origin of tho solar system

An increasing complexity of form and increase of mass characterize the expansive pressure of life throughout time on the planet Earth.

atmospheric oxygen reacted with iron to make rust and with other metals such as uranium to make metallic oxides. But, eventually, over millions of years, the oxygen rose to poisonous levels. Some of the bacteria worked their way into the ground to avoid it but others adapted, learning how to use the oxygen in a process called *respiration*. Today almost all animal cells host an oxygen-using organelle called mitochondria. Lynn Margulis, an American microbiologist, hypothesized that in time bacteria teamed up to form larger organisms which evolved into amoebas, bread molds, toad stools and, eventually, higher plants and animals such as elephants and humans.

Margulis worked together with James Lovelock to demonstrate the mechanisms by which the biosphere regulates itself. Lovelock and Margulis showed in detail how the biosphere could operate as a *cybernetic* system — *cyber* meaning helmsman — a system which is self-regulating by means of rapid microbial response to small changes in the composition of the atmosphere.

Margulis dramatically developed this point of view in *Microcosmos*, describing the change effected by microbial populations of an early biosphere when confronted by the pollution of their environment with their own by-product, oxygen. In relatively short order, a great number of microbes had evolved to become aerobic — using oxygen in their respiration — and the threat to the continued life of the previously anaerobic biosphere was resolved. The cybernetic process does not have to be thought of in terms of willful or conscious decision-making:

> *"Microbes apparently did not plan to bring under control a pollution crisis of amazingly daunting proportions. Yet they did what no governmental agency or bureaucracy on Earth today could ever do. Growing, mutating, and trading genes, some bacteria producing oxygen and others removing it, they maintained the oxygen balance of an entire planet."*

New Horizons

The "space age" that was heralded in 1957 with Sputnik, and was well underway by 1969 when men landed on the moon, had opened the way out of the biosphere, outward towards the planets and stars. And people saw the haunting face of a mysterious planet silhouetted against the vast expanse of empty space — the Earth.

Buckminster Fuller coined the term 'Spaceship Earth' in the same year that man first walked on the Moon, noting that:

> *"... omission of the instruction book on how to operate and maintain Spaceship Earth and its complex life-supporting and regenerating systems has forced man to retrospectively discover just what his most important forward capabilities are. His intellect had to discover itself. Intellect in turn had to compound the facts of his experience ... objective employment of those generalized principles in rearranging the physical resources of environment seems to be leading to humanity's eventually total success and readiness to cope with far vaster problems of the universe."*

The prospect of leaving Earth for long voyages or permanent settlements made it necessary to consider what is usually taken for granted: air to breathe, water to drink, food to eat, shelter from excessive solar radiation, and maintenance of a safe range of temperatures. The best human engineers could provide and control these factors for short durations. What entity had accomplished them in the biosphere so long? Buckminster Fuller, and G.E. Hutchinson's philosophical musings rang true — humankind should consider itself as being on the verge of emerging from its infancy, having been sustained by a generous world tolerant of humanity's mistakes and ignorance.

And humans were awakening to their circumstance just as the situation was becoming dangerous, in part due to their biological success. Twenty-five hundred years ago there were only fifty million human beings (0.05 billion) scattered over the face of the Earth. A thousand years later, by sixteen hundred, the human population reached its first billion individuals. Only three hundred years later, by nineteen hundred, there were two billion. The third, fourth and fifth billion marks have all been passed within the last eighty-five years.

While early humans did occasionally pollute and exhaust a locality with agricultural practices, the damage was local — though sometimes long lasting. The ancient city of Uruk, home of the first chronicled hero of humanity, Gilgamesh, was brought to ruin by the civilization he established. The hero, Gilgamesh, King of Uruk, set out to carve his name in history by conquering nature, and cut down the sacred trees of the forest in the thrall of technical power. The barren desert that resulted destroyed Uruk, and remains to this day!

Our tool-making and thinking species progressed from poking holes in the earth with sticks, to strip mining, chemical factories, vast urban sprawls, and even unleashing the atomic forces within matter itself. The specter of a fiery war to end all wars — and perhaps all higher forms of life on Earth — was born in a mushroom cloud. The lyrical sound of the syllables Hi-ro-shi-ma is a synonym for a hell that no one wants to

Space Shuttle Columbia on launch from Kennedy Space Center Florida (NASA).

know. With the bomb the technosphere had evolved to a point where trial and error are no longer acceptable. Knowledge of the biosphere was urgently needed to understand and control the effects of man's now significant impact.

Increasing population and the scale of human activity left fewer corners into which mistakes could be swept and ignored. Moreover, the ability to measure and track environmental damage kept pace with industrial and agricultural technology. James Lovelock had invented an *electron capture detector* in the late 50s, a more precise refinement of the gas chromatograph, which measures the presence of chemicals in extremely small amounts. Nuclear testing and pesticide residues tracked throughout the food chain on Earth with the aid of this detector verified the thesis of naturalist Rachel Carson's *Silent Spring*, which in the 60s warned that global circulations of air and water were distributing pollution far from the sites of its origin.

By the 1970s, no one needed gas chromatographs to know there were subtle and dangerous chemicals loose in the environment. Major world cities gasped for breath. Garbage was 'thrown out' only to reappear on beaches or in one's harbor, like a tar baby that could not be shaken off. Taller and taller smoke stacks only sent the air pollution elsewhere — it came down as acid rain, destroying forests and lakes.

View from NASA Space shuttle. In order to stay in space, enclosed recycling evolutionary life systems or biospheres will be needed.

NASA

The idea of the biosphere, thinking of the system as a whole, began to be discussed in more and more quarters. The name and emphasis varied — the biosphere, Gaia, Spaceship Earth, the global commons, the global environment — but the fundamental idea was the same. Wherever humankind wanted to go, whatever we wanted to do, we were going nowhere—and wouldn't survive — without a biosphere. And the spaceship Earth, as Buckminster Fuller eloquently pointed out, was a boat we were all in together.

In 1972, the British economist Barbara Ward collaborated with Rene Dubos in preparing a report on the global environment for the United Nations, *Only One Earth: The Care and Maintenance of a Small Planet*. The problems, they wrote, were unprecedented and so were the solutions needed:

"The whole human way of life is, as it were, pulling at its anchors in nature and in history and straining to set sail. Or should we rather say that it is gathering energy on its launching pad to take off, rocketlike, into regions as relatively uncharted as the surface of Mars?"

Whether those uncharted regions held riches or disaster were in the hands of humanity:

"... the two worlds of man, the biosphere of his inheritance and the technosphere of his creation, are out of balance, indeed, potentially in deep conflict. And man is in the middle. This is the hinge of history at which we stand. The door of the future opening onto a crisis more sudden, more global, more inescapable, more bewildering than any ever encountered by the human species. And one which will take decisive shape within the life-span of children who are already born. No problem is insoluble in the creation of a balanced and conserving planet, save humanity itself. Can it reach in time the vision of joint survival? Can its inescapable physical interdependence, the chief new insight of our century, induce that vision? We do not know. We have the duty to hope."

In the mid 60s, Stewart Brand, editor of the *Whole Earth Catalog*, knew that the space agency NASA must have photographs of the Earth taken from spaceships, and publicly asked: "Why haven't we seen a photograph of the whole Earth yet?" Brand had buttons made which asked precisely this question and began selling them across America. NASA released the first photograph of the planet in 1967, and Brand published it on the cover of his *Whole Earth Catalog*. He felt that the image did much to shatter both the 'flat Earth' and 'endlessly more to be used' illusions. Seeing the planet as a whole supported the idea of thinking of it as a total system and solving the problems at hand on the right scale. Like it or not, humankind had changed the face of the planet and had to acknowledge responsibility. "We are as gods," he wrote in his 1980 edition of the catalog, "so we might as well get good at it."

At the conference initiating the Biosphere 2 project, astronaut Rusty Schweickart recalled his spacewalk during the Apollo 9 flight when he looked on the Earth, miles below: "On that small blue and white planet below is everything that means anything to you; all of history and music and poetry and art, death and birth, love, tears, joy, games ... all on that little spot in the cosmos. National boundaries and human artifacts no longer seem real. Only the biosphere, whole and home of life."

NASA

Earth from space. "The biosphere, whole and home of life". And from this "Biosphere 1" must come the other biospheres that will travel throughout space.

Early Experiments

The vision of a biological future in space was prophesied in the late nineteenth century by Russian scientist Konstantin Tsiolkovsky, who wrote about the principles of rocketry, spaceflight, and "space greenhouses" which he recognized would be necessary for living in space. Tsiolkovsky became an inspiration and touchstone for the Soviet research in closed ecological systems which began over three decades ago, headed up by Evgenii Shepelev and Ganna Meleshka, both scientists at the Institute of Biomedical Problems, Moscow, directed by Oleg Gazenko. When their first experiment was written up a quote from Tsiolkovsky served as the introduction:

> *"It is impossible to exist for a long time in a rocket: the supply of oxygen for breathing and food will soon run out, the byproducts of breathing and cooking contaminate the air. The specifics of living are necessary — safety, light, the desired temperature, renewable oxygen, a constant flow of food ..."*

ROBERT HAHN

In 1961, the space age was barely four years old. Evgenii Shepelev led the Moscow Institute of Biomedical Problems laboratory team in designing 'closed loop' biological systems to support cosmonauts in space. Challenging biomedical problems—especially adaptation to extreme conditions of cold, heat and altitude — were a special field of interest. Shepelev, Meleska, and their chief, Gazenko, were interested in the practical unification of Tsiolkovsky's theories of space as a cosmic *future* and Vernadsky's theory of life as a cosmic *energy source*.

Shepelev, Gazenko, Gitelson, the three men in the center from left to right, visit Biosphere 2 in May, 1989. John Allen is showing the rice production paddies. These three Russian space scientists were pioneers in closed-loop biological systems.

Shepelev had constructed a steel cylinder with a volume of one hundred and sixty cubic feet — about what you would have in a cube that was five and a half feet on each side — completely sealed and permitting virtually no air exchange with the outside world. He proposed to enter the small chamber with only one other living species as an ally: the green algae, *chlorella*. He had calculated that in this chamber he and a tank with less than eight gallons of the algae in a water solution could suffice as an absolute minimum for a closed self-sustaining system. The algae would take in carbon dioxide and give off oxygen as it photosynthesized. His respiration would supply the mirror image of the equation as he breathed in oxygen and gave off carbon dioxide. In theory, the two *should* balance. Shepelev crawled into his steel chamber and sealed the door behind him. Twenty-four hours later, the living proof of the theory opened the door of the test chamber and walked out of his experiment. But, from the look on his colleagues' faces, he could tell something was wrong. After a few moments outside, he returned to the test chamber and breathed in a lungful of the air inside. It stunk! The chlorella system was too simple to cleanse the air of the many trace gases which humans, plants and steel chambers give off. Evgenii Shepelev silently wondered how he had managed to last out the day-long experiment. In addition to failing to recycle the complete range of atmospheric gases, the productivity of the algae declined over the first twenty-four hours and began to fail completely during the second day of operation. The astonishing thing was that such a simple system had worked at all.

The research team led by Shepelev with his co-worker Meleshka learned how to maintain the algae's productivity by adjusting the nutrients in the water tank. They also upgraded the tiny apparatus, affectionately nicknamed the *Siren*, to a more sophisticated little habitat. Soon they were ready to try a month-long test with a human occupant.

Although Shepelev's experiments with artificially created closed ecological systems were the first, the work of Professor Clair Folsome also made a major contribution to understanding biospheric principles. Folsome headed a laboratory of microbiology at the University of Hawaii, and had written on the origin of life.

No one had ever thought to try the simple experiment which he began one day in 1968. Taking several one and two liter glass flasks from his laboratory, he walked out to the sandy beach of the Pacific Ocean and scooped some sea water into them. He sealed the ocean water — complete with its natural diverse complement of microbes — inside the flask so that no air or nutrients could enter or leave. Bringing them back to his laboratory, he set them on the windowsill in indirect sunlight, and left them to their own devices.

A surprising thing happened. They lived. And lived and lived ... and lived. Clair respected the tenacity and ingenuity of microbial communities, but even he had not expected this result. He gathered more samples from different locations and the collection of flasks on his windowsill expanded. He discovered that, given a high enough diversity in the microbial populations — and only artificially cultured mono-cultures did not have this necessary diversity — the little ecosystems lived indefinitely. They bubbled away in healthy biological activity: photosynthesizing, feeding, converting wastes back into usable forms, doing their 'microbial thing.' Folsome developed methods to study them without violating closure, and began a sophisticated analysis of their properties. Since the little ecosystems were separate from the biological cycles of the external environment outside the flask, they were free to set new standards. And they did, with generally higher rates of cycling and higher oxygen levels than the global biosphere.

For a long time the general reaction of many scientists to Folsome's life-in-a-bottle experiments was: so what? Few seemed to grasp the far-reaching implications of Folsome's *ecospheres*. It was nearly ten years before he was able to publish his first paper on the subject in a scientific journal. Folsome combined his two major interests, space travel and the origin of life on Earth, to become an expert in extraterrestrial biology. His Laboratory of Exobiology became an important research center, one of the first to study rocks brought back from the Moon.

Folsome's work with closed life systems grew out of his strong belief that humans would soon be living and working in space. Lynn Margulis

They lived ... and lived ... "The oldest closed ecological system assembled by the late Clair Folsome in 1968. Its atmosphere differs significantly from Earth's biosphere and is maintained by its inhabitants in a narrow range.

KENT WOOD

CARL SHANEFF

Portrait of Clair Folsome holding an ecosphere. Folsome was on the Review Board of Biosphere 2 until his unexpected death, and contributed greatly to understanding the functions microbes perform.

introduced Folsome to Space Biospheres Ventures at the outset of the Biosphere 2 project. "He's the one who can really help you," she said to John Allen and Mark Nelson. Until his untimely death at fifty-three, Folsome was a member of the Review Committee for the Biosphere 2 project — a vast scaling-up from the glass flasks that sat on his laboratory windowsill. But the small flasks, from one-half liter to five liters were the precursors to a life-system under glass that will encompass more than seven million cubic feet.

Before starting to build Biosphere 2, Allen and Margret Augustine from Space Biospheres Ventures made a journey to Folsome's laboratory and pored over the meticulous data which showed that each of these tiny worlds had made and stabilized their own unique atmosphere within certain definite limits. This was a striking exemplar of the Margulis-Lovelock mechanism theory, and provided essential information for planning the building of Biosphere 2. It is a fitting tribute to Folsome's convictions that most of his tiny worlds under glass are still viable, including some that date from that first stroll to the seashore.

The year was 1972. Soviet research in closed ecological systems begun by Shepelev was continued at the Institute of Biophysics under the direction of Josepf Gitelson. Few people outside the soviet space program would know of the historic trial being prepared at the Institute's facility in remote northern Siberia, near the city of Krasnoyarsk. The *Bios* facility that Gitelson and his team had assembled was another self-contained system. This one, bigger yet and containing higher plants as well as algae tanks, was the most advanced in the world.

Life in the Shepelev's *Siren* had been a bit of an endurance test not only because of its limited life support capabilities, but also by its small size which limited activities — comparable perhaps to the situation of an astronaut during the early launches. The United States had made its own attempts during this time at developing the algae-based life support systems, running up against similar limitations in function and design.

Bios-3 was about the size of a small space station, three hundred and fifteen cubic meters in volume, and the residents stepped through the airlock into a remarkably complete little urban world, with an human habitat and small 'garden.' The lives of the people inside were remarkably full. They tended their garden and algae, harvested the crops, prepared the food they had grown in their kitchen — even baked their own bread. They could watch television, and read the newspapers, books or mail which were delivered via an airlock so as not to break the atmospheric seal — this was before the era of electronic computer mail and compact videos. They also kept in touch with the outside world visually, through a glass porthole, and by telephone.

The space age farmers of the Bios experiments raised crops under artificial lights in liquid fields — hydroponic farming using solutions of water containing the essential minerals which plants need to grow. Their results indicated that life support for one person might be possible in about two hundred and thirty square feet of growing area.

The Bios-1, Bios-2, and Bios-3 experiments from the mid-sixties to 1984 were landmark steps in the development of successful closed ecological systems. The Bios team attained closures of up to six months with a three person crew that remained healthy and alert and with virtually one hundred percent of the air and ninety-five percent of the water recycled by the plants which produced half the food. However, the human fecal wastes and some inedible plant materials were removed from the system and necessary proteins, including meat, had to be brought in.

These experiments showed the ultimate limitations of closed non-ecological systems: essential trace elements were lost or became unavailable after a time because of the removal of solid human waste and plant material. Gases emitted by the algae tanks proved toxic to the plants and there were problems with the air quality as certain trace organic gases built up over time. The soil-less agricultural system provided too few habitats for the microbes which do so much in their invisible worlds to keep the biosphere in balance.

In November 1986, Space Biospheres Ventures made a special two week trip to Moscow to meet Gazenko, Shepelev and Gitelson. It became clear that Shepelev and Gitelson's work had made major breakthroughs in enclosed life systems that included humans and shown that they could survive for limited periods in truncated ecologies. Folsome had proved that a full suite of microbes could construct and maintain a world. The next step seemed to require the development of a full-fledged ecosystem approach to closed systems. Allen, Nelson, Augustine and their colleagues would make full use of the Bios and Folsome data in designing their own closed system experiments.

"The Space Age Farmers of Bio-3". Their vegetables were grown in liquid solutions. Bios-3 recycled all of its own air, ninety-five percent of its water, fifty percent of its food, but little of its waste.

The Idea of Biosphere 2

"If you don't do it, then who will?" Coming from Buckminster Fuller, this was not a rhetorical question. It was a challenge. Fuller whacked his wooden pointer repeatedly across the podium and stared piercingly around the room at the eighty participants gathered at *Les Marronniers*, a conference center in the heart of Provençe, France. Called *The Galactic Conference*, the 1982 conference of the Institute of Ecotechnics covered a galactic range of subjects with no holds barred. This conference was the seventh in a series of conferences the Institute held over the period of a decade inviting leading scientists, explorers, and artists to attend. The first years focused on different biomic regions of the planet, then the planet Earth itself, followed by larger systems such as our solar system, the galaxy, ending with *The Cosmos Conference* in 1983.

MARIE ALLEN

Above: Buckminster Fuller and Bill Dempster discussing the theory and practice of building high strength to weight ratio structures.

Right: Lynn Margulis, one of the leading thinkers on biospherics, addressing the Institute of Ecotechnics on the key role of microbes.

MARIE ALLEN

Other speakers at the conference included microbiologist Lynn Margulis who presented her theory about life on Earth. "Human beings are not special, apart, or alone," Margulis asserted. Humans need to learn to live in a new way, she suggested, because life on Earth cannot be sustained by technology alone. "Only the photosynthesizers can harvest the sun."

Architect Phil Hawes came to the podium carrying a white sphere the size of a basketball. "Why not look at life in space as *a life* instead of merely travel?" he said. "Why not build a spaceship like the one we've been traveling on — along with all its inhabitants?" He held up the globe. He called it *Galactica*. He spun it around. The other side was cut away to reveal an elaborately detailed three-dimensional interior. Inside were gardens, housing units, a jogging track, research laboratories, and a pool beneath a waterfall. Hawes proposed that cramped modules designed to speed passengers from here to there were a thing of the past, that the well-designed space habitat of the future should, like Earth, produce its own atmospheric conditions and have miniature farms, Japanese tea gardens, and wilderness areas.

MARIE ALLEN

Phil Hawes presenting a possible space colony life system geometry at the 1982 Institute of Ecotechnics' Conference.

Sitting in the audience, Institute directors Mark Nelson and John Allen were intrigued by the effect of the architectural model on this audience of thinkers and scientists. Hawes' model had been the first public presentation of their ideas. After the conference they sat talking, sketching, and speculating even further on integrated ecological systems. With engineer Bill Dempster, they calculated the biomass of various plants that would be necessary to generate a specified volume of oxygen. If you started with a small enclosure, say seven to twenty thousand cubic feet, and put one person inside, how many plants would you need to use up carbon dioxide and make enough oxygen? The faces of others sitting with them registered surprise. "Why, it might actually work" someone exclaimed. "Of course it would work," Allen grinned, "if we got the right size. What do you think the biosphere does every day?"

The key conceptual breakthrough realized by those Institute members gathered to work on this was that *biomes* indeed function as the main building blocks of the biosphere, and that the driving and adjusting forces were the microbes working with the gases in the atmosphere. They began to integrate these concepts with all the previous work to date, and concurred that a model of the biosphere would not be complete without a diversity of biomes and microbial suites, as represented in the Earth's biosphere.

Obviously, the arctic tundra and the deep ocean were not practical. Instead, the biomes of the tropical belt, that region of life girdling the equator and among Earth's most productive, were selected — tropical rainforest, savannah, desert, and ocean, along with two "man-made" biomes — agriculture and city.

One of the most complex tasks would be to establish and maintain the different biomes in close proximity to one another and to supply the range of functional microbial suites necessary. If these could be constructed and maintained within a closed system, then a claim could conceivably be made to having built a *second biosphere*. What to call it? Earth was the first biosphere. This would be ... Biosphere 2!

GILL C. KENNY

GILL C. KENNY

GILL C. KENNY

John Allen (top) Director of Research and Development; Margret Augustine (middle) C.E.O. and co-designer; Mark Nelson (bottom) Director of Space Applications; Edward Bass (inset) Chairman of the Board.

The People

The initiating group that decided in 1983 to actually build Biosphere 2 was known as the "Decisions Team". All of its eight members — John Allen, Mark Nelson, Margret Augustine, Edward P. Bass, William Dempster, Marie Allen, Kathelin Hoffman, and Robert Hahn — had worked together on a number of complex projects since 1974, and three of them since 1967. The first problem to deal with was that the building of the biosphere was estimated to cost thirty million dollars — not counting securing the location, performing the scientific and engineering research, designing and architecting the structure and training the biospherians. The only capital Decisions Team had at that point was intellectual, moral, and fifteen years of project experience.

COURTESY OF EDWARD P. BASS

That same year, Edward Perry Bass, well known in the business circles of Texas, suggested that Decisions Team make a joint venture with his venture capital firm, Decisions Investment. The joint venture would be called *Space Biospheres Ventures (SBV)* to indicate the threefold nature of the enterprise. Ed, also a Director of the Institute of Ecotechnics, and personally conversant with the ecology and engineering aspects of the project, became Chairman of the Board.

It was almost a foregone conclusion that the President of SBV and Chief Executive Officer would be Margret Augustine, Managing Director of *Synergetic Architecture and Biotechnic Design (SARBID)*, based in London. Together with her colleague, Philip Hawes, a former student of Frank Lloyd Wright and a colleague of Bruce Goff, they were to design and architect Biosphere 2. Augustine and Hawes had worked together before on the design and construction of a sailing ship, a hotel in Kathmandu, and a condominium complex in Santa Fe. Margret was nearing completion on the *Caravan of Dreams* cultural arts center for the commercial heart of Fort Worth, Texas when the Biosphere 2 project was presented to her. Responsibility for the engineering design would be placed in the hands of William Dempster, whose background ranged from computer programming at Lawrence Radiation Lab in Berkeley, to engineering and quality control. John Allen, a founding director of the Institute of Ecotechnics, with a background in geology, ecology, engineering, and Harvard Business School management, would occupy the post of Executive Chairman the first two years while Augustine, as Project Director, got all the first contracts together, beefed-up the architectural corporation, and upgraded the site. Allen would later become Director of Research and Development. Mark Nelson, Chairman of the Institute of Ecotechnics, was put in charge of the Space Applications. These five had worked together before on innovative projects in remote areas, but this one would challenge all of their combined skills and experiences.

The Place

The choice of site for Biosphere 2 was crucial. The search soon narrowed to the Southwest because of the climate, and thence to southern Arizona, which proved to be rich in ecological, scientific/ technical, and agricultural expertise as well. For example, there was Tony Burgess, Peter Warshall, both outstanding ecologists, and Carl Hodges with his Environmental Research Lab at the University of Arizona. The Institute knew them all well and considered them among the leaders in ecological and environmental science. By a great deal of effort and remarkable good fortune, a site was found near Tucson that seemed ideal. The site had more than two thousand acres of open land and, while there had been no "for sale" sign posted, the owner was willing to sell.

Tony Burgess agreed to work with the project on desert ecology, Peter Warshall on the savannah ecology, and Carl Hodges to help with the agricultural and engineering systems to insure that the total system heat would stay within acceptable limits. The surrounding communities were home to skilled workers, artists, and professionals, a good source of excellent staff as the project developed.

At nearly four thousand feet above sea-level, the site also promised to take the edge off the burning summer temperatures and still had clear air at those times when the low-lands were covered with haze and pollution. The scenery was breathtaking. To the delight of SBV, the site came equipped with the facilities of the former Motorola Executive Training Center, which made it possible to welcome visiting scientists and hold conferences straight away, as well as provide housing for a nucleus of the managerial and scientific staff. By July 1984, Space Biospheres Ventures had moved in. Biosphere 2 had a home; it was christened *SunSpace Ranch*.

Sunspace Ranch, located overlooking the Canyon del Oro, facing the Catalina Mountains, north of Tucson, Arizona.

D. P. SNYDER

19

2
The Quantum Leap

"Extremely interesting looking out into that other world, under different metabolic laws; so unmeasured as to be primitive in comparison, unknowing as to the air they breathe, the build of elements/molecules in their medium. 'They know not what they do —' but now can no longer be forgiven this.

We will move again in the air-water ocean like the great whales, but this time in co-regulated harmony. And our co-regulated air-water oceans will meiosis time and again, populating space and time, building up improbable concentrations of energy until the secrets of 'big-banging' and 'black holes' are fully revealed.

> *What a multi leveled dyad!*
> *from Cosmology to Freud*
> *from Physics to Metaphysics.*
> *Between the Big Bang*
> *and the Black Holes*
> *Arise a hundred billion galaxies*
> *Each with a hundred billion stars*
> *Each with a hundred billion biospheres*
> *Each with a hundred billion dreams*
> *Each with a hundred billion bits*
> *All facing death*
> *All facing the one-in-a-million*
> *Long shot for transcendence*
> *The long roam throughout*
> *the Unknowable*
> *the Unbeable*
> *the Ununderstandable*
> *but not Uncontactable*
> *Until impossibility itself*
> *Becomes a local option*
> *And biospheres co-regulate*
> *the Universe,*
> *the Cosmos,*
> *the Absolute*
> *And we stream in and out*
> *of sunlit local gases*
> *on our sensors of delight*
> *at the same time collecting and recollecting Memory*
> *for that shipwrecked moment*
> *when we need it all and then some*
> *to keep the voyage going going*
> *and never gone."*

Vertebrate X

The Test Module

T he first requirement for Biosphere 2 was that it be completely sealed — sealed so that virtually no exchanges of air would occur once its door was closed. Shepelev had accomplished this in his steel chamber. Gitelson's Bios-3 was large enough to support two or three crew members, and produce about half their food, but suffered a leakage rate of about fifty percent a year. Since it was a test for the micro-gravity conditions of space flight around planet Earth where direct penetration of the Sun's rays would not be possible, its energy for photosynthesis came from high-intensity sodium artificial lights powered by energy from the outside.

For the same reason, artificial light had also been used in NASA's remarkable CELSS (Controlled Ecological Life Support Systems) program, launched in 1977. Among the researchers involved in CELSS, Frank Salisbury and his team at Utah State University had boosted plant densities for wheat from the 200-300 plants per square meter usual in field plantings up to 2,000 per square meter and were aiming for up to 10,000 per square meter. Salisbury anticipated that a farm the size of an American football field "...would support 100 inhabitants of Lunar City".

Folsome's small ecospheres were closed and let in sunlight, but upscaling a glass globe to proportions necessary for even a single human habitation, much less construction on Mars, was unfeasible.

Biosphere 2 Test Module system schematic. Surface area for plant growing area includes mezzanine platforms and areas within hollows of the space frame.

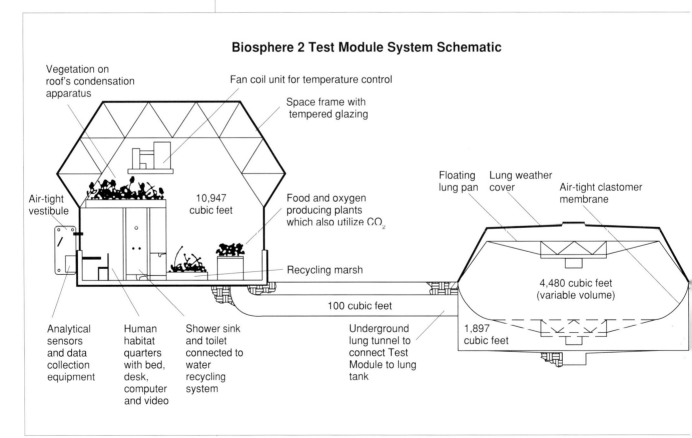

Biosphere 2 Test Module System Schematic

Vegetation on roof's condensation apparatus

Fan coil unit for temperature control

Space frame with tempered glazing

Air-tight vestibule

10,947 cubic feet

Food and oxygen producing plants which also utilize CO_2

Floating lung pan

Lung weather cover

Air-tight clastomer membrane

4,480 cubic feet (variable volume)

Recycling marsh

100 cubic feet

1,897 cubic feet

Analytical sensors and data collection equipment

Human habitat quarters with bed, desk, computer and video

Shower sink and toilet connected to water recycling system

Underground lung tunnel to connect Test Module to lung tank

The creators of Biosphere 2 had a quantum leap in mind. The initial plan was just under three acres — two million cubic feet of space — about the size of three enclosed football fields with fifty foot ceilings. Biosphere 2 planners not only wanted to perfect a closed system, like Shepelev's, and let in plenty of light, like Folsome's; they wanted to make it hundreds of times larger than any previous closed ecosystem made, far larger than Bios-3. They wished to make a model as close as possible to the Earth itelf. They wanted to set up an experiment that could determine the laws of biospherics by including in their model a sufficient variety of biomes, atmosphere, soils, and rocks, to make a homologue of Biosphere 1. If the model worked, they would be a step further along in understanding those laws.

"Human nature and science as a whole are reductionist in nature," said Walter Adey, Director of the Marine Systems Laboratory at the Smithsonian Institution, and one of the biomic consultants to Biosphere 2. "When you look out at the biological world, it looks unbelievably complex and you try to simplify it. And just about everything we do with nature is along that vein of simplification and I think in many cases we create new problems for ourselves in so doing." By chopping biological systems into small pieces, scientists may be destroying basic self-controlling and buffering mechanisms and actually making some aspects of them *more* difficult rather than easier to understand. "I concur that we should include as much diversity as we possibly can. I think we'll all be surprised at the stability that we derive from it."

This approach, to put as many elements in as possible and let "natural forces" achieve the balance, would not eliminate the possibilities of problems even with a structure as large as Biosphere 2. Atmospheric cycles that recirculate water on Earth wouldn't occur in the confines of a building, even a large one. Biosphere 2 would need mechanical pumps to circulate water; it would need fans to circulate air; it would need motor-powered devices to simulate waves. And it would need plenty of diversified biomass for the atmospheric cycles to interact with. Adey had expressed the essential point: Biosphere 2 must be built according to the laws of life, and a synergy with the laws of engineering technology would have to evolve.

Biosphere 2 was moving into uncharted areas hoping to find out about processes that are not yet understood even in the ecosystems of Biosphere 1. It was not clear what would happen in a much smaller, sealed, sunlight driven Biosphere 2. It *was* clear that experimental knowledge had to be increased at a rapid, even prodigious, rate in order to succeed. The next step was to build a prototype apparatus large enough for one person to live in, with an ecosystem based on the same principles of Biosphere 2.

So, the *Test Module* was conceived.

From Folsome's ecospheres to the Test Module would be a jump of some five orders of magnitude, a "scaling-up" by a factor of 100,000. From the Test Module to Biosphere 2, the jump would be around three orders of magnitude, a scaling-up by a factor of a thousand. The size of Biosphere 2 had been calculated by taking into account best advice on the minimal possible size of the various biomes. If the surface area of the tropical belt of Biosphere 1 were compared with that of Biosphere 2, there would be a scaling-down of some ten orders of magnitude:

The Test Module is the first building one meets when entering SBV's Biospheric Research and Development Center.

Biosphere 2 could be thought of as ten billion times smaller than Biosphere 1. Scaling-up or scaling-down is the supreme test of the engineer, the crucial quantitative-qualitative point.

The size of the Test Module was determined primarily by the need to give life-support to one human being. In early 1985, William Fred Dempster, the chief systems engineer, and John Allen were furiously thinking about the size to build the Test Module when Margret Augustine made a key managerial observation: "We not only need to test the life sciences," she said, "we need to test the space frames, the sealing, the building contractors, the SBV staff, the agricultural systems engineers, and the computer people, all at the same time. We're going to have to scale up at each of these points."

It was probably the single best move made on the way to building Biosphere 2: to make the Test Module a testing device for the *entire* project, not only for closed life systems. Margret could observe not only the progress and problems in the science of closed systems, but also the morale, ability, teamwork, and cost-effectiveness of each component of her overall team. She designed the Test Module to be of a critical size to make a modular test of each main element: science, engineering, business, communications, training, and architecture. There was now a clearcut two-stage mission: make the Test Module work; make Biosphere 2 work. Two scale steps between the one-liter ecosphere and Biosphere 1 would be filled in.

The First Tests

In January 1987 the first unmanned experiments were begun in the Test Module — at last a hands-on opportunity to learn how life responded in closed systems. The series of experiments, which ran from January to May, marked the first time that soil-based systems had been used in a closed ecological system. Inside were savannah grasses, rainforest shrubs, small trees, desert cacti and a sampling of agricultural crops — a range of plants from all of the biomic areas of Biosphere 2 — to gauge their response to the high humidity and comparable light levels they would experience in the Biosphere. The plants grew rapidly, some spectacularly. Savannah grass plants about to seed were included to test if pollination would be impaired by the high moisture content and the lack of breezes. The soils had been inoculated with selected bacteria and fungi; insects had been put in; sensors were installed to monitor air and water.

JEFF TOPPING

Linda Leigh checking the effects on the plant growth following one of the early Test Module closures.

Excitement was great as people gathered in the analytical lab to read the daily and nightly sensor readings. Carbon dioxide levels dropped during the course of the day as the plants fixed it in their photosynthesis; at night they rose steadily as the microbes and plants shifted into oxidation mode and breathed out the carbon dioxide as waste product. There was a delighted recognition that here was a vibrantly responsive ecosystem living inside the Module. Observers could literally read off the 'pulse' of the system by watching the numbers change.

The day of the first re-opening of the system was memorable. It had been shut for a week. Would there be a repeat of Shepelev's air problems? The opening was scheduled down to the last minute: the taking of air samples, water samples, soil samples, the insect and plant observations. The news that crackled back through the walkie-talkies told it all: "Marvelous smell. The air is absolutely marvelous."

It was clear that the system worked. No massive die-offs had occurred. No microbe, insect, or plant had run destructively rampant. No "green slime" climbed the glass walls. The larger plants, set in pots on the upper level of the Module, had doubled in size. Allen, one of the first researchers to walk in after the seal had been broken, remarked that the pungency of its atmosphere brought up memories of "lying face down in the rich plowed Oklahoma bottomland in early May." Soil samples were sent to Clair Folsome at his laboratory in Hawaii. He analyzed them and found that all the major functional microbial types were present and in good quantity.

Not only was progress being made with the ecological systems, it was also being made on the major engineering problem to be solved for Biosphere 2: "Making sure the roof doesn't leak," as Margret Augustine succinctly put it. New methods of glazing and new types of sealant - termite proof! - were being rapidly developed and tested, in an effort to meet the goal of reducing leakage to only one percent per year. This would mean the air would turnover only once in a century.

Below: John Allen, Test Module "Vertebrate X Experiment", September 10-12, 1988. The first human in a closed ecological system experiment where one hundred percent of the air, water, food, and waste were recycled.

Bottom: Waste recycling systems used in the Test Module.

The Human Experiments

With about two years till the scheduled closure of Biosphere 2, the first human closure in the Test Module was planned. John Allen volunteered to be "Vertebrate X", the first in what would become a series of experiments involving humans in the ecosystem. Messages heralding the event were sent around the world to pioneers in the field. Shepelev replied that: "Man is the most unstable element in the ecosystem". When asked for a word of advice, he said: "Courage!"

ROBERT HAHN

A three-day closure was set for September 10, 1988. Allen would be the first man to live in a *completely* closed ecological system: one hundred percent of his air would be recycled as would all of his water; all his food would be grown inside, and all of his wastes recycled.

The Test Module was fitted with an air purifier called a soil bed reactor that used the microbes in the soil to assist in preventing gaseous toxins from accumulating. An overhead second tier was constructed to fit in more savannah grasses. A small marine area had been added. Water for drinking would be condensed from the humid air inside the Module. Tests had indicated that water would be of the highest purity, having already been passed through the plant leaves. Food would be supplied by a miniature tropical garden which included pineapple, peanuts, tomatoes, lettuce, potatoes, sweet potatoes, dill, beets, carrots, and onions. A rice and tilapia fish system would supply high protein foods. Mint and lemon grass would furnish tea.

This was the first test of the waste recycling systems. Phil Hawes had established contact with Bill Wolverton of NASA Stennis Center who advised on the design of a marsh plant recycling system for all the human wastes. They would pass through an anaerobic (oxygen-less) phase, then be circulated in a tank with aquatic plants where microbial and other aerobic decomposition would occur. Then they would pass to a lagoon system prior to being used in the irrigation systems for the plants. All this would happen in an area of less than twenty square feet.

A new element in the Test Module was the human habitat. It was outfitted with simple stove and kitchen, lavatory, shower, bed, dining table and desk. One of the questions to be answered was: would the gas emanating from the materials these things were made of be cleansed by the soil bed reactor?

ROBERT HAHN

ROBERT HAHN

There was a medical team for the experiment, backed up by Dan Levinson, M.D., of the University of Arizona Medical School. Allen would have an outside crew on duty twenty-four hours a day, with "watches" changing every four hours, as on a ship. The full computer system hadn't made it in time! A blackboard outside his window would carry current readings as well as graphs and predicted levels. He would maintain telephone contact with the outside team. For medical safety he wore an alligator-clip on his finger when he slept which read both pulse and the level of dissolved oxygen in his bloodstream. Since this was the first total human enclosure no chances would be taken. For privacy at night he could draw the blinds for sleep, but the electronic readout would be visible to the watch crew members on duty.

The analytic laboratory would also be in operation round the clock. Not only would carbon dioxide, oxygen and seven key trace gases be monitored continuously by sensors, but samples of air would be taken every half hour and tested. The quality of both potable and waste water would also be monitored.

TOM WIEDWANDT

Left: John Allen measuring pulse and oxygen in his blood during the "Vertebrate X Experiment".

BERNT ZABEL

These three photographs show the analysts monitoring the key trace gases and taking manual samples to check their instruments.

TOM WEIDWANDT

TOM WEIDWANDT

C. ALLAN MORGAN

While all this data was being gathered and people got a taste of excitement at seeing a person actively participating in the life support system that maintained him, Allen was able to reflect on being simultaneously the experiment, the experimenter and the experiencer of a different ecology. The journal he kept gives us a glimpse into this world. It reveals that his senses seem heightened, unusually attuned to everything around him. He would picture the microbes in the soil decomposing toxins and emitting carbon dioxide, realizing that this meant more work for the plants. He was keenly aware of the plants: sugar cane, rice, water hyacinth, the sweet potato vine curling along a ledge ten feet above his head. The air he sniffed was moist, heavy, seemed saturated with oxygen, carrying the clean scent of lemon grass and mint. He soon became used to the low hum of the fan, the only sound except for voices over the telephone that he heard during seventy-two hours. A grin broke across his craggy face. He imagined he was on another planet.

Like the two women who would later go into the Test Module for longer stays, Allen felt more keenly than ever the unity among living things.

September 10, 1988, 19:05

"Already a strange partnership seems to be building between my body and the plants. I find my fingers stroking, feeling the soft rubbery texture of the spider plant, knowing it's picking up outgassing products, the dark masses of the plants beyond the room perhaps turning off their last benedictions of oxygen, the C_4 savannah grasses above perhaps still turning out some in the last lingering light. The microbes in the soil bed also sorting out the toxins, decomposing them to edibles, but also emitting CO_2."

September 11, 1988, 09:25

"CO_2 staying 2500 +. Cloudy day. Notice my attention turning more and more to condition of plants. How's their water? Soil? Have to stay in good shape to produce the O_2, suck up the toxins. I've always had the sense of plants being alive, responsive, even a living symbol, but now they're necessary. Of course I knew that before, but I didn't really sense/feel their necessity. And since they're necessary, I look out for them."

ROBERT HAHN

John Allen with his journal and reference books during evening in the "Vertebrate X Experiment".

Towards the end of his stay, Allen was feeling his health improve and began to feel more and more that conditions inside the Test Module were far better, in many important respects, than those outside!

"My attention's moving on to sun and water now and the soil. I don't mean directed attention ... I mean a fascinated attention now lurks very near the surface of consciousness and finer and finer changes of sensation radiating from the sun, water, soil, as well as plants now possess the power to put the organism into an alert, investigative state."

September 12, 1988, 18:15

"It appears we are getting close to equilibrium, the plants, soil, water, sun, night and me. Inside an hermetic seal, in communication by word and sight (ear and eye) with the world, but touch, taste, and smell all different."

At the end of the three days, Allen emerged healthy, glowing and feeling positively mystical. He said: "I knew with my body as well as my intellect and emotions that Darwin and Vernadsky were right about the power of the force of life."

In March of the following year, 1989, the second human enclosure took place. This time it was for five days. "Vertebrate Y", Abigail Alling, said afterwards:

"I realized when I came out that the physiology of my organism went through a lot of change. During the five days my physiology had changed to the point where I felt I was experiencing the system in a different way than when I had first gone in. I felt, for example, that the moisture, the humidity, was very much more a part of my skin, of my physiology, that I was more receptive to receiving moisture. I hardly drank at all for example ... When I came out, into the desert, I could feel my physiology clamping down..."

Alling felt that the Test Module had not come to equilibrium:

"I felt it change. I felt that the system was responding to my addition and myself responding to the ecosystem. All the elements in the atmosphere had leveled off except for CO_2 which continued to rise, but its rate of increase was falling, indicating a reaching of equilibrium well before reaching critical levels."

Abigail Alling, now the Associate Director of Development, made the second experiment as "Vertebrate Y" for 5 days in the Test Module.

30

TOM WEIDWANDT

Left: Linda Leigh, now the Associate Director of Research, made the third and decisive experiment in the Test Module, "Vertebrate Z", for 21 days. Here she consults with Mark Van Thillo on the plumbing system.

Abigail Alling coming out of "Vertebrate Y Experiment", healthy, alert and enthusiastic.

This was critical. The next human enclosure, of "Vertebrate Z" Linda Leigh, would be for three whole weeks, time enough to show whether the system could settle down. The experiment which ran in November 1989 acheived an equilibrium state. The results from that experiment provided convincing evidence that Biosphere 2 was do-able. Like the other vertebrates, Leigh found herself keenly aware of the effects of her own actions:

> *"If I would dig my sweet potatoes for the week, I would be disturbing the soil which creates more carbon dioxide in the atmosphere. And I could tell that at the end of the day on my graph of the CO_2 that I had increased it by my digging. So, I would try to organize my harvesting based on when there was most light, or sometime during the day when it made sense in terms of carbon dioxide."*

During her stay, Leigh swapped experiences with "Buzz" Aldrin, the astronaut, linked by video hookup to the support facilities outside. She told him:

> *"I could visualize a different terrain on the outside and I was pretending that maybe it was a Martian or lunar terrain."*

She was quite convinced that the system could support somebody *indefinitely.* These experiments were signicant steps forward — a successful closure involving humans meant a quantum leap across the unknowns to Biosphere 2.

31

3
Biospherics

"I was thinking this globe big enough until there sprang out so noiseless around me myriads of other globes."

Walt Whitman

Biospherics is Born

Biosphere 1, cosmically powered, 3.8 billion years old, two trillion living tons networked through thirty million species — contains our past, present and destiny. Biosphere 2, scientific model, symbol, affirmation, helps us understand life and, therefore, ourselves.

The first Biospheres Conference convened in December 1984 at Sunspace Ranch to deepen the discussion and development of Biosphere 2.

Biosphere 2 would be open to energy inputs: sunlight would provide the basic energy for photosynthesis and electrical energy for technical systems. Heating and cooling would be generated outside the Biosphere and delivered to the inside. Just as Earth sends and receives radio waves, light beams, and communications to and from vehicles in space, the Biosphere would also be open to information exchange. A significant purpose of the experiment would be to create a dialogue between the biosphere of Earth and another biosphere and make comparisons which had never before been possible.

Allen began to list the biomes now planned for the system: a human habitat, an agricultural area, a rainforest, a savannah, a desert, a freshwater and a saltwater marsh.

"And an ocean!" Hawes called out from the audience.

"Oh, yeah. Right. An ocean," Allen answered with a grin.

He wrote it down on the blackboard with the others. No one really knew how they'd pull that off. "At that point no one had a clue about how to build an ocean," Phil Hawes later recalled.

Ultimately, it seemed impossible to make anything modelled on Biosphere 1 without an ocean. It is not clear that a biosphere in near space whould have to have an ocean to operate successfully, but it was firmly resolved that Biosphere 2 should have one.

SBV's vision of a *created biosphere* was 180 degrees from what they were thinking about at NASA. The space agency's idea of a space station was of high-tech "cans" containing lots of plastic, metal, computers, and walls painted in psychologically approved colors. There was little room for the odors, textures, sights, and sounds of a complex living assemblage of plants, animals, and bacteria. Yet this is what the people of SBV saw when they dreamed of large stations in space or settlements on other planets. Was the plan too radical? Was it "do-able"?

The Biosphere 2 ocean was to be composed only of high yield components: beach, lagoon, coral reef and adjacent ocean in order to compensate for its relative lack of area compared to Biosphere 1's ocean.

Opposite: Light is as fundamentally essential to Biosphere 2 as to Biosphere 1.

It became clear during the three days of the conference that there was a huge communications gap between the engineers and the ecologists. The root of the problem was the difference in attitude of the two camps. Ecology was — and to a considerable degree still is — a rather "fuzzy" science. Ecologists have historically been more accustomed to dealing with relationships and energy paths than with columns of numbers. Engineers, on the other hand, usually deal in the numerical world of well understood laws governing the behavior of materials. They're trained to design for the worst possible case, for ten percent contingencies and to make things idiot-proof. It was not too surprising that each group at the conference sometimes found the other to be baffling and at times infuriating. By the end of the conference it was clear that if Biosphere 2 was ever to be built — and work — the two groups would have to find a way to listen to and understand each other.

The unifying power of the Biosphere vision was such that they did. A dialogue began to develop. In design meetings and through the electronic networks, ecologists began to ask the questions for which they might need engineering answers. They began to quantify things they qualitatively understood. Engineers began to see consequences to life relationships when they added or subtracted a quantity.

A lot of talking went on in 1985 — in person, in small workshops, over the phones, over electronic network —as individuals in the two groups started to work together. They began to learn — and develop — *biospherics*.

The two man-made (anthro-pogenic) biomes, agriculture and "city", can be seen on the left, and the five wilderness biomes, rain-forest, savannah, desert, marsh and ocean on the right.

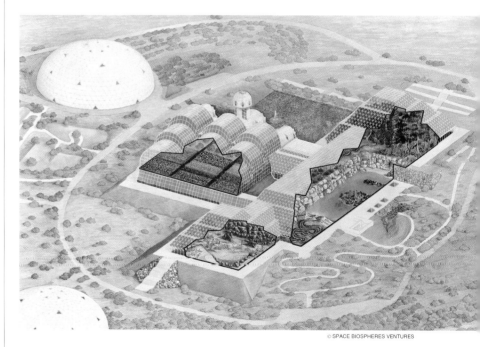

Minimum sizes began to be assigned to the different biomes — a half acre each went to agriculture, rainforest, and desert, a fourth of an acre to savannah and ocean, an eight to the estrarine marsh. This in itself was a remarkable agreement among the consulting design ecologists who grew to include Ghillean Prance on rainforest biome design, then of New York Botanical Garden, now Director of Royal Botanical Gardens at Kew, and Walter Adey, Director of the Smithsonian Marine System Lab on ocean biome design, as well as Burgess, Warshall, and Hodges.

It was agreed that the life scientists didn't have to meet the mechanical scientists unless necessary, nor the energy scientists, nor the communication scientists. Each group designed its own "dreamworld," and it didn't have to change anything *unless* it made one of the other worlds not work. In other words, each group had veto power — ecologists over technologists, or vice versa, but no one group could tell another what to do. And the reasons for any veto had to be spelled out to make certain that such changes were not for the sake of change, ideology, or excessive caution.

Meanwhile, conflict began to rise among different factions within the life sciences group. Agricultural experts rarely work with rainforest botanists, who in turn don't often have reason to talk with marine biologists. Crop specialists worried about invasions of insects from the wilderness biomes into the agricultural area. Entomologists had nightmares about 'their' insects being completely consumed by the birds and reptiles that would be picked by the vertebrate experts. "How many moths does a bat really eat?" became a burning question. The rainforest people worried about the effects on their plants of salt spray from the ocean. "Can't we forget the ocean?" The marsh folk were concerned about minerals that might leach from the rainforest soil and get carried into their biome.

The savannah and desert shared a thornscrub boundary. Who would design that? Workers in the agricultural biome campaigned for artificial lights to make food production more efficient. Eloquence and hard data flowed in defense of each position, a process which shed a great deal of light on details known to specialists in each discipline but unknown to others. Biosphere 2 needed to work as a system or none of the individual parts would work at all.

Ecology

Legend tells that Noah admitted each kind of animal, two by two, to his ark which sheltered them until the deluge had passed. Details of diet and cooperation between the animals during their stay in the ark, however, remain a mystery enshrouded by time and myth.

Unlike the ark, Biosphere 2 welcomed animals to a *web of life,* as participants in the *oikos logos*. From the Greek origins *oikos* (house) and *logos* (governing rules), ecology literally means the "rules of the house".

The rules of the house for Biosphere 2 were much the same as for ecosystems of the Earth. Residents had to earn their keep, performing some useful function in the ecosystem. All the work of the ecosystem had to get done — all the functional niches must be filled. A candidate's behavior must be in keeping with its co-inhabitants and environment. In turn, the ecosystem must have the resources and an appropriate environment to support its residents through all phases of their lives. Biosphere 2 was similar to an island ecosystem in that the residents could not leave, and new populations or individuals could not migrate in to rejuvenate or replace resident populations.

For all the differences — like lower ultraviolet light because of the surrounding glass sky, somewhat lower levels of sunlight due to the spaceframe structure, and the fact that the edges of the 'island' were to be absolute — the house was similar enough to the biosphere of the Earth to permit the biome design team to use known principles to construct models of the food web before any animal or plant was actually introduced.

To eat and be eaten is the fate of all creatures. In the words of microbiologist Clair Folsome, "Life is an ecological property, and an individual property for only a fleeting moment." But what about the human species? What would be their role in Biosphere 2? That of the *keystone predator*, meaning that predator without whose efforts a given trophic chain in a given ecosystem would suffer from one or more devastating population explosions. Skilled in the art and science of naturalist observation, the biospherians could act as predator to whatever level of the trophic chain was beginning to make undue depredations upon the biomass or species diversity of a given biome. By the same token, the rest of the Biosphere must supply the biospherians with pure air, clean water, tasty, abundant, and nutritious food, and a variety of stimulating impressions and scenes.

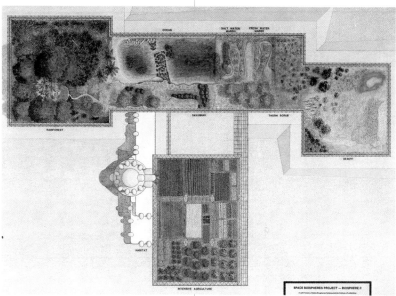

The floor plan view of the seven biomes shown in the previous cutaway.

Once the number of people, a crew of eight, had been agreed upon, it was possible to make rough estimates for atmospheric volume and overall areas of the biomes because required food production and atmospheric needs could be calculated. On the atmosphere two schools of thought immediately formed: one advocated essentially two different atmospheres, one for the anthropogenic or man-made biomes, i.e. city and agriculture, and another for the "wilderness areas", terrestrial and marine. Biospherians would communicate between the two by a special airlock hatch; in this way the agriculture and city regions could be protected against invasions of unwanted populations of microbes, fungi, and "bugs", and the wilderness area could be guarded against CO_2 buildup and other possible pollutants. The other school critiqued this plan on the grounds that this would mean Biosphere 2 would no longer be based on Biosphere 1 and, indeed, would not be a biosphere at all, but rather a closed highly engineered life support system in one portion, and a romantic, nostalgic biosphere of forty thousand years ago (before modern man) in the other portion. The anthropogenic biomes would only be a more cumbersome though expanded replay of previously done work, and the wilderness biomes would waste money better devoted to maintaining some of the little virgin wilderness left on planet Earth.

It was decided to make the atmosphere unreservedly one, to make the physical passages between the anthropogenic biomes and the wilderness biomes clear, but with some means to control bio-damaging interchanges of bacteria, fungi, plants, and insects.

Ocean and Marsh

As the time approached for the second Biospheres Conference in 1985, questions about scale mounted. Arguing that two thirds of the Earth's surface is water, marine consultant Walter Adey lobbied for more surface area of water. Or, if not more surface, then a deeper ocean. In the end, finances helped to decide the depth of the ocean at twenty-five feet; going ten feet deeper would have cost an additional ten million dollars. The ocean would cover approximately fifteen percent of the total surface area of the wilderness section of Biosphere 2 and contain a coral reef and coastal lagoon. But, proportionally, it would be ecologically more productive than Earth's oceans because it would exclude the low-biomass "water deserts" of mid-ocean and deep water areas beyond the reach of sunlight. The ocean project generated so much enthusiasm, in fact, that Biosphere planners included in their plans an underwater viewing gallery for visitors. And feeding into the ocean on a tidal flux would be another of Walter Adey's dreams: the marsh system would be modelled after the Florida Everglades.

In cooperation with the National Park, an Everglades Expedition was mounted in July 1988 to bring back the requisite subsystems.

ROBERT HAHN

ROBERT HAHN

By 1985, Adey had built several *mesocosms*, intermediate sized living ecological demonstration models, at the Smithsonian Natural History Museum in Washington, D.C. He had thought about creating an estuarine marsh modeled after the Chesapeake Bay, but was sure the estuary wouldn't be viable on the scale of his previous projects. It needed a greater size, and Adey had neither money nor space for it. A chance encounter put him in touch with people who wanted a large-scale transplantation of a coral reef. It was that summer, when one of Adey's graduate students returned from a conference sponsored by the Space Studies Institute in Princeton, N.J., and told him about a man named Phil Hawes and a project called Biosphere 2.

Adey thought it sounded like Biosphere 2 could use his mesocosms. He talked to Hawes and soon was making trips to Arizona. Nelson invited Adey to the September 1985 Biospheres Workshop, following which Adey consented to work as consultant to SBV providing the expertise for the Biosphere 2 ocean, its marsh, and streams. SBV agreed to finance

Walter Adey of the Smithsonian Marine Systems Laboratory had worked out a concept with SBV and soon had a working mesocosm in Washington

C. ALLAN MORGAN

ROBERT HAHN

JEFF TOPPING

Top left: The Biosphere 2 beach has its own complex ecosystem.

Above: Walter Adey on the Everglades Expedition, July 1988.

Top right: Biosphere 2 ocean biome by September 1990 was operating as a living ocean, holding its nitrate composition, for example, to a small fluctuation around ten parts per billion.

his building of the Chesapeake Bay and Florida Everglades mesocosms at the Smithsonian as testing modules for Biosphere 2. Teaming with him from SBV was Abigail Alling, who would see that the task of getting an ocean and marsh system to Arizona was accomplished.

Adey found a home for the Chesapeake Bay mesocosm in the basement of the Smithsonian and built it much like the marsh that would exist in Biosphere 2, except with artificial lighting. It consisted of eight interconnected tanks filled with water and sludgy marshland soil, supporting fish, insects, bacteria and plants. The salinity of the water shifts gradually but inevitably from fresh at one end of the mesocosm to sea salt at the other. Simple but effective "tide machines" raise and lower the water level in the mesocosm, and send pulses of water back and forth. Augustine and Allen visited it four times and had long lunches with Adey before they all became convinced that this method would actually work. On the fourth visit, Adey said, "Smell the basement air! It's just like Chesapeake Bay!"

GILL C. KENNY

Left: The Biosphere 2 marsh biome was a living system by early summer of 1990.

Below: The marsh starts with its fresh water pond and becomes progressively saltier through the three different mangrove systems

D. P. SNYDER

Having lots of nutrients in water is okay for a marsh, but around coral reefs nutrient levels are extremely low. High nutrients will literally poison them. As a solution to the problem, Adey had developed algae scrubbers, a system that mechanically passed water through populations of algae, to let the plants clean the water. These would operate in a room off the savannah cliff face of Biosphere 2 under artificial lights. An algae scrubber is a fairly simple machine made of a plastic trough in which a mat of algae grows. As water flows over and through the mat, the algae feeds on the organic waste released into the seawater by the coral. The algae scrubbers thus perform the same function for their coral reef that algae and other photosynthetic life forms carry out in the wild. In Biosphere 2, the ocean and marsh water would circulate through a system of sixty algae scrubbers. Each is equipped with a flush bucket which would tip when full and send a flow of water down the length of the algae-covered mats. Besides cleaning the water, the scrubbers aerate it, increasing its oxygen content the way waves and tides do in the ocean.

Rainforest

When the Biosphere 2 project was being organized, British-born Ghillean Prance was the first and foremost candidate for rainforest biome design consultant.

> *"After so much experience with the destruction of the rainforest, I was excited to work on the construction, the building of a rainforest. Why would I, as a rainforest scientist get involved in building half an acre of rainforest in Arizona? Many factors of forest conservation and the contribution of the rainforest to the biosphere are based upon vague generalizations. The fact is that there is much we simply do not know. In a closed environment, we can set up all sorts of tests and make measurements which are impossible in the open greenhouses of a conservatory, or in the open system of the rainforest itself. Biosphere 2 was a wonderful opportunity to create a situation for interactions between rainforest plants and to begin to understand them — to create new interactions and perhaps new solutions. Could we make it as good or better than a half acre of rainforest of the Chacobo Indians? Can we make it even more productive, utilizing all we are learning about the functioning of the rainforest? Can we use it as a demonstration of how the despoiled and abandoned areas of Amazonia and the other rainforest can be treated? Can we use it as a showcase to show the alternative to destruction?"*

Prance is now Director of the Royal Botanic Gardens at Kew, United Kingdom, but when he first met members of the Institute of Ecotechnics in 1979 in Peruvian Amazonia, he was the Director of the Institute of Economic Botany at the New York Botanical Garden. Prance spent over twenty years working in the Amazon region as New York Botanical Garden's Vice President of Research, coordinating pioneering work in flora surveys, ecological relationships and documenting the traditional Amazonian Indian's use of rainforest species.

COURTESY OF GHILLEAN PRANCE

Above: Ghillean Prance, the Director of Kew Gardens in the United Kingdom, has for years been completely at home in rainforests.

Right: Orinoco, Venezuela. This was the location for the SBV rainforest collection expedition in early 1989, arranged in cooperation with the Venezuelan government and the New York Botanical Garden's Institute of Economic Botany, founded by Prance.

HARRY SCOTT

Nature had tens of millions of years to perfect the way of life of the rainforest, and to evolve plants and creatures of nearly unimaginable diversity in form and function to bring their own special flourish to the biosphere. Indeed, most ecologists think the rainforests are the greatest storehouse of genetic inventiveness on Earth, representing the inherited 'natural intelligence programs' which life developed to overcome environmental problems, cleverly use an available opportunity or produce valuable chemicals. Tropical plants are the source of the medicinally active compounds in a quarter of the prescriptions filled in American pharmacies — including birth control pills, blood pressure and heart regulating medications, and the drugs which have lifted the nearly inexorable death sentence formerly associated with certain forms of childhood leukemia. Half of all modern medicines contain natural products. Many of the rainforest's contributions to the biosphere are so integrated into modern life that their tropical origins are forgotten: rubber, coffee, chocolate, vanilla, cola, pepper, bananas, papayas, pineapple, coconut, rattan cane, to name merely a few. The majority of beautiful ornamental plants so common to American and European homes evolved in the shade of the tropical rainforest floor.

The Biosphere 2 rainforest biome in September 1990 shows the reproduction of the key features of the environment for the "lost worlds" of the cloud forests.

41

*Right: A closer view of the
Biosphere 2 rainforest cloud
plateau.*

*Below: Another photo from the
Orinoco Expedition.*

HARRY SCOTT

C. ALAN MORGAN

The building of a rainforest in Biosphere 2 would be a chance to conduct a long-term study of a rainforest of minimum critical size, minimum critical size being the smallest diversity of species with smallest populations within a species that can still be considered a viable rainforest community. Creating a rainforest was a challenge of urgent scientific value on Ghillean's agenda. "Humankind has become expert at making deserts out of rainforest," Prance says, "but we have yet to prove ourselves adept at the reverse process."

Savannah

"*S*avannahs are the benediction of the Earth, the original grasslands that allowed human beings to become the first pastoral tribes on the planet. In physical anthropology, it is now thought that the whole human body evolved from leaving the forest, where it had gotten to be a strong, upright form by hanging from trees. Over time, the body became too big and when they came down from the trees in a stooped kind of way, they walked out onto the savannah. Eventually—because it was good to walk in the savannah—man became a walking creature with an upright body. The savannahs have a crucial place in human history, for it was an unfilled niche of the savannah that allowed human evolution to occur.*"

This was said by Peter Warshall, anthropologist, ecologist, and design consultant for the savannah biome of Biosphere 2. In designing the savannah he recommended the creation of a 'synthetic ecology' rather than the analog approach used in the rainforest and some other biomes, thus presenting the opportunity to leave out undesirable elements from particular ecologies and take outstanding elements from widely separated geographic areas. Together, these elements have the genetic potential to produce a super-savannah biome because originally they were together in the super-continent, *Pangea*, before continental drift separated them, each to evolve a similar savannah, but each in its own special way. The savannah biome of Biosphere 2 includes ingredients from South America, Africa and Australia.

Having evolved with conditions of disturbance, the savannah requires a little "kick" now and then. And not just a *little* according to Warshall, but a lot. "The savannah is in fact so adapted to being disturbed that it cannot sustain itself without it. The savannah is a very resilient community with wide ranging tolerances." There was talk at one point of putting up a sign in the savannah saying "Please Disturb".

Peter Warshall, collecting grasses in Guyana with the cooperation of the government, for the Biosphere 2 savannah.

NEIL RETTIG

The savannah can have forests, but they are thorny and do not have the three tiers of the wetter rainforest.

PETER WARSHALL

43

JEFF TOPPING

NEIL RETTIG

Above: There are forty-five species of savannah grasses in Biosphere 2.

Opposite: The stream meanders through Biosphere 2 savannah.

Left: A photo of the grassier sections of a savannah region in Biosphere 1.

Desert

There is a desert for every rainforest, if you take a global view, and it's clear why. The Sun warms the equator of the Earth, and warm humid air rises and fans out, some going north, some south. The air cools as it rises, and the condensed water vapor falls as rain on the rainforests. The planetary winds continue in their migration away from the equator, not only with no more rain to bestow, but actually drawing out moisture from the landscape below on their return toward the equator. Thus, as David Attenborough writes in *The Living Planet*:

> *"Wherever land straddles the equator, there are pairs of deserts, north and south. The Sahara is matched, south of the rain soaked forests of central Africa, by the Kalahari and the Namib. The Mojave and Sonoran deserts in the southwest of the United States have their equivalent in the Atacama in South America. And in Asia, the vast deserts of Turkestan and central India are paralleled ... by the great deserts of central Australia."*

Deserts are one of the most recently evolved biomes on the planet. Small areas of aridity and life forms adapted to them may be as ancient as any land species of the planet. But, the spread of extensive desert conditions, however, is a geologically modern development of perhaps the past five million years. It is possible to trace the evolution of plant species from moister biomes into forms able to take advantage of the expanding desertic realm.

Tony Burgess was known to the staff of SBV as the man responsible for designing with the Institute of Ecotechnics the Desert Dome exhibit built on the roof of the *The Caravan of Dreams* performing arts center in the heart of Fort Worth. He was the natural choice as biome design consultant for the desert of Biosphere 2. Burgess served as botanist for the Eco-Hydrology Project of the United States Geographical Survey (USGS) and has conducted extensive fieldwork in the Southwest, Northern Mexico and Baja, California. According to him: "Deserts are in a state of permanent disequilibrium. Their organisms must switch frequently between frantically exploiting rainfall and stubbornly surviving drought."

Natural deserts are an integral feature of Biosphere 1 and their strengths, needed in any model of the biosphere, quickly showed their advantages for Biosphere 2. "No liquid water leaves the desert, so it is the only place where soluble salts can accumulate. The temperature tolerance of the ecosystems is greater, both seasonally and daily, and the desert is prepared to go into dormancy if water is not available. We could complement the growth cycles of the other biomes — the desert could be dormant in the summer when other biomes were rapidly growing and using carbon dioxide, and have its growing period in the winter when the biosphere needed plants to take up carbon dioxide to keep the atmosphere balanced. There is even a complete range of natural deserts on Earth from which to choose our model for Biosphere 2. People ask, how will you keep it dry in there with that ocean? The answer of course is that we will do what nature does — simply work with available conditions, in this case, high humidity. Luckily, there is also an ocean not too far from Tucson, and on its shores — a fog desert!"

C. ALLAN MORGAN

Tony Burgess studying the progress of the savannah in Biosphere 2, May 1990.

JEFF TOPPING

LINDA LEIGH

Above: The cool, fog desert of Baja California which, thanks to the cooperation of the Mexican government and ecologists, furnished a good deal of the genetic material for Biosphere 2's desert.

Left: The desert in Biosphere 2, thanks to the skill of Tony Burgess and Linda Leigh, reproduces the qualities of Biosphere 1's cool foggy deserts: Namibia, Baja, Chilean and Southern Arabian peninsula.

LINDA LEIGH

JEFF TOPPING

Above: Another photo from Baja California.

Above right: And a comparable area in Biosphere 2.

Diversity

An early challenge of designing the wildernesses of Biosphere 2 was figuring out how to put a tropical forest, a savannah, an ocean, a marsh, and a desert under one canopy. In nature such diverse types of terrain are separated by miles of transitional ecosystems. In Biosphere 2, the savannah, traditionally called a *transition zone*, was right between the desert and the rainforest.

The presence of an ocean, marsh, small ponds, and streams, sources of constant evaporation, insured that Biosphere 2 would have sufficient humidity. The tropical rainforest, in fact, required lots of moisture for its existence. Warshall planned a savannah which could live with that by choosing plants from humid savannahs, although that somewhat increased the threat of fungal infections. Warshall still had to deal with the prospect of the termites, nematodes, and mites that populate the savannah soil making a getaway to the rainforest or the desert. But basically he knew his savannah could live with the other areas.

The *boundaries* between biomes had to be designed with particular care. For example, the typical plants of rainforest take none too kindly to seawater spray from an ocean. In Biosphere 2, a special copse of tall bamboo provides shelter against the drift of salt through the air. Then, in transition from savannah to desert, a thornscrub slope fits very nicely. In the marshlands, the fresh water part is protected against any tidal backflow from the saltier regions by a difference in the water levels, just as in Biosphere 1.

The designers of Biosphere 2 were pushing for maximum diversity with many alternative food pathways, the recipe for ecological stability. That meant having lots of different species. The question was whether to include species in numbers that they were certain would survive or jam a much larger number together and let natural extinctions occur — Biosphere 2's own version of survival of the fittest. Lack of genetic variability due to small populations could doom some species to an extinction that they have avoided in the larger world.

Prance favored overloading the rainforest and letting the species fight it out. Some species, however, were indispensable. If they were lost, the integrity of the entire rainforest would be at risk.

He developed several rationales for choosing the over two hundred higher plant species of the rainforest. Some would fill specific ecological niches. Others played prominent roles in the succession of plants as the community approached climax. He also needed canopy plants, ground cover, shrubs, epiphytes (airplants) — such as orchids — and vines. For special human treats and needs, cocoa plants could produce chocolate and others would provide fruit, medicines, fibers, and flavors.

To create a real jungle would take time. It wouldn't be possible to produce a mature rainforest by the start of the two-year closure. Rainforest trees take years to reach maturity and form a canopy even in a greenhouse. Prance planned to follow the natural process that occurs when a forest clearing has been abandoned by growing various fast-growing secondary forest and colonizer species to provide the shade and structure for the slower growing species and lower canopies.

The first years of designing Biosphere 2 focused on figuring out how to create the physical environments for the different biomes — to make the rainforest warm, to have waves in the ocean, to clean the air. By July 1986, discussions of the living contents of each of the biomes were in full swing. During one session, Ghillean Prance was shouted down when he suggested including howler monkeys in his tropical rainforest. Others who had also visited the Amazon reminded him of the beast's propensity for howling and throwing things. He conceded, but still, he insisted that he'd find some interesting animals. And he — and others — did!

GILL C. KENNY

An overview of the cliff edge that divides the ocean from the savannah in Biosphere 2, October 1990.

4
The Household

"Shovel in the microbes, shovel in the seeds, shovel in the biospherians and close the doors," one scientist told Leigh.

Others wanted precision. "Let's quantify every microbe," another said, "Let's measure everything that we're putting in to the Nth degree."

One contingent wanted a whale.

Members of the Household

According to the legends of the Pima Indians of the Sonoran Desert, a great flood came and forced the people to take shelter in a large boat. After many days of floating, Spider Woman sent a hummingbird aloft to discover if the deluge had receded. The hummingbird returned with a flower in its beak, a sign that they would soon safely disembark. Then the hummingbird flew off and gathered some clay to take to Tcuwut Makai, the Supreme Being. With the clay, Tcuwut Makai made more people to repopulate the Earth. The flood waters will return, so the legend goes, if those children of the hummingbird ever come to harm.

On that basis alone hummingbirds seemed good candidates for Biosphere 2. But those selecting the hundreds of species for this unprecedented experiment also had other concerns. Hummingbirds are prodigious pollinators. In their search for the nectar that fuels their high-speed metabolism, they visit many flowers and pick up and distribute pollen, the male germplasm for the sexual reproduction of plants. Thus the hummingbirds assure the reproduction of the plants whose flowers provide their food.

The highest point of Biosphere 2's space frame is only ninety feet tall, about forty feet above the cloudforest mountaintop. Hummingbirds that flew higher than that in their mating dance would smash into the glass "sky," running the risk of killing themselves. Peter Warshall, working with Linda Leigh, SBV's Terrestrial Biome Coordinator, was in charge of selecting candidates for the 'household', and he needed a low-flying species without a loop-dee-loop mating ritual.

Evolution has produced an array of beaks, some extremely long, giving those birds exclusive reign over flowers with long tube-shaped blossoms. Others curve to the left or right and up or down, matching a twist in a particular type of blossom. They narrowed the list to species with modest, straight beaks, and favored those "generalist" species known for their broad tastes in flowers, not gourmets demanding an exclusive brand. The hummingbird chosen would have a medium to long bill, enabling it to feed on longer-barreled flowers whose nectar was out of reach of bees and butterflies.

Flowers and nectar are not the only source of food for some hummingbirds. Some species have an appetite for insects. The entomologists demanded that these be removed from consideration in favor of those that eat insects only occasionally.

Warshall wanted to pair the hummingbirds with at least one other

GILL C. KENNY

How many flowers does it take to keep a hummingbird alive?

Opposite: The final hummingird selected for Biosphere 2 has had to pass a wide variety of tests before receiving its ticket for the first two year sojourn. This is a hummingbird from Biosphere 1.

species that would do at night what the birds did during the day. The obvious candidate was a bat, one that would occupy the same ecological niche as the hummingbirds but at different times of the day.

Finally, when the process of elimination had cut the list from three hundred and forty hummingbirds to about five, he chose the sparkling violet-eared hummingbird, a hardy version easily obtainable. There was one drawback. For some weeks out of the year, Biosphere 2 may have no flowers for the hummingbirds to feed upon. The only solution was to plan to provide supplemental food for them — the biospherians would have to feed the birds. Despite their lengthy roster of duties, no one complained.

The question of bats came up on a cold, sunny morning in February of 1990. Leigh and Warshall sat down and started weighing pros and cons, just as they and other biome captains had done before and would do again. This was part of a continuous process of picking the thirty-eight hundred species for Biosphere 2 from among the thirty million now estimated to live on Earth. Warshall and Leigh laid out a grid of significant points for the bats, including their ecological function, public interest that would encourage education in ecology and biospherics, source of food, requirements for nests or protected areas, and the cost of buying them.

The number of bats selected might be the deciding factor. Donna Howell, a bat specialist who works for The Nature Conservancy, had recommended a minimum of eight to ten. Bats are colony creatures and need to have others around them when they roost during the day. Before making a final decision, they decided to see if they could lower the number to five or six.

Next on the list was the pygmy kingfisher bird. The kingfisher, which feeds on fish, hard-bodied insects such as beetles, frogs, and small lizards, would be Biosphere 2's top predator, if chosen. No other animal would come close to its appetite.

A nectar feeding bat pollinating an agave in the desert.

© 1991 C. ALLAN MORGAN

For the first two years, the biospherians would probably have to feed them a part of their diet. Otherwise they might do great damage. The populations of lizards, fish, frogs, and insects needed time to establish themselves, to grow, reproduce, and flourish. Turtles eat fish and lizards eat insects. Vine and garter snakes also eat lizards. The kingfisher would compete with all of these and might seriously deplete the numbers. If the kingfishers were to be on the inside, the biospherians would have to feed them a commercial bird chow to keep their predation down. On the other hand, the kingfishers would be a potent evolutionary force, keeping their prey species "on their toes".

Selection Process

Anticipating the effects of one species on another, and then on all the others, was the constant theme of the selection process. As part of his work on picking vertebrate animals for the wilderness biomes, Warshall would draw a sketch of the structure and list around it some of the species that would go in. These species maps were a valuable tool, presenting in graphic form the names and locations of different animals. Next to the name of some of the animals he put an arrow that extended out to the right. The arrow's length indicated how far afield the animal might roam and in which biomes it might live.

Next to the slug-eating snake and the red-footed tortoise Warshall put a question mark. The doubt had to do with the complicated tasks of figuring out what a species ate, what ate it, whether it could adapt to the confines of Biosphere 2, and numerous other factors. Among reptiles definitely "in" were several species of anoles — or tree lizards — such as the knight anole which puffs out a rose-colored dewlap. The anole's uncanny habit of turning and rotating its head to different angles gives it a pensive look that has been interpreted as signs of its moods. The black-neck garter snake, the blue-tongued skink, the leaf/tree frog, the parachute gecko, the prehensile-tailed skink, and a species of tropical leptodactylid frog — a group known for distinctive toes — were naturals.

The bizarre matamata turtle would take more deliberation. It's an odd creature: with a flat, triangular head ending in a flexible snorkel; a long, broad flat neck from which shaggy lobes of skin hang; and three longitudinal rows of large humps running over its flat shell. The turtle eats by snapping its mouth open so quickly that a vacuum sucks in nearby fish. At night time in the wilderness, biospherians walking around with flashlights would see its eyes, which are lined with a crystalline layer that reflects light like a crocodile. Shortly before closure it still had not received an invitation, as the scientists awaited answers to questions about the availability of fish for it to eat and the micro-habitat in which it resides. The slender vine snake was also in doubt, as it competed for the same sized lizards as the kingfisher.

On Warshall's chart, an arrow from the anoles extended off to the right, as far as the lower savannah. These lizards could be expected to live and eat not only in the rainforest, but to roam into the savannah biome and possibly into Burgess's fearsome thicket of thornscrub. Arrows next to the vine and garter snakes showed that they would range as far as the desert biome at the south end.

In the savannah, reservations were made for the neotropical wood turtle, the granite night lizard, two more species of anoles from the southeastern United States, another species of leptodactylid frog, the canyon tree frog, leaf-toed geckos, brown tree lizards from the southwestern United States and northern Mexico adept at disguising themselves as branches, and the Texas tortoise. The granite night lizards and another anole, the *anolis sagrei*, were cliff-dwellers and would live in the cliffs that separated the rainforest from the savannah. The canyon tree frog would make its home around the waterfall coming down from the rainforest and in the thornscrub forest ecosystem.

Some distinguished citizens of Biosphere 2:

C. ALLAN MORGAN

Blue tongued skink.

C. ALLAN MORGAN

Yellow footed tortoise.

JOHN CANCALOSI

Spiney lizard.

More leading citizens:

Cuban Anole.

Green tree frog.

Galagos.

Beside the spiny bufo, a toad native to the Sonoran Desert and the fossorial snake, Warshall had scribbled question marks. The leaf-toed gecko and the Texas tortoise could be expected to range into the thornscrub and the desert biome. The tree lizards would make their home in the thornscrub.

Warshall, Leigh, Howard Lawler, and the other animal consultants had several reptile species tagged for the thornscrub or the desert, including the side-blotched lizard common to dry areas of the American West; the desert night lizard, which is one of few lizards to bear young by placenta birth rather than hatched eggs; and the coleonyx, or banded gecko, the unblinking reptiles that clean their large eyes by licking them with their tongues. The spadefoot toad and the red-spotted toad were possibilities for both the desert and the thornscrub. About thirty side-blotched lizards would be included, ten males and twenty females.

Only one reptile was slotted for the marsh biome, and that was the terrapin. Others would probably migrate over from the pond area; some tiny inhabitants had come in with the flora itself, namely a few tiny crabs and frogs. Warshall had originally picked about eight reptilian species for the freshwater pond, but had cut that list to six after discovering that the pond would be smaller than he had expected. The southern toad and the striped mud turtle were dropped. Still on his list for the pond — mere finalists at that point — another anole, the brown snake, the green tree frog, the peninsula cooter turtle, and the ring-neck snake.

These larger, active vertebrates, like the bats and kingfishers, were difficult to meld into Biosphere 2. Not only did their appetites threaten to overwhelm other species, but, unlike plants, they had to have space to move about. Chopping down a few plants wouldn't be so bad. But if the wrong species of animals were put in and it became clear it would devastate the ecosystem, the biospherians would be faced with living with its destruction, hand-feeding it, or killing it. The naturalists wanted to avoid that kind of decision. But they also wanted to include a species that would capture human interest — the bushbaby was a natural candidate.

Bushbabies, also known as *galagos*, are prosimians, monkey-like mammals. They are quite small and nocturnal in their habits. Leigh and Warshall figured that Biosphere 2 only had room for two or three of them. But their petite cuteness, big eyes, and playful ways gave them the potential for stardom. Since cats and dogs were out of the question, bushbabies would fulfill a role as companions for the eight humans. Bushbabies are omnivores, feeding on fruits, insects, and gums from trees. They would fill no special ecological function unfilled by other species. In fact, Leigh and Warshall were not even sure which niche they would eventually occupy. They would find themselves in extensive and intense competition within the food web. So, like the kingfishers, the biospherians could end up feeding them at times during the first two years, although they would get first crack at the fruits produced by the gingerbelt. Bushbabies move around by walking through tree canopies and vines, so they'd need a network to travel through the wilderness. They also crave contact with their kind — or the closest thing to it. They may well adopt the humans for playmates.

Insects

Scott Miller, Head of the Entomology Department of the Bishop Museum in Hawaii and consultant to SBV, set down the criteria for selecting the insects in Biosphere 2. First, those like the termites which play a crucial role in the nutrient cycle. Second, those which, like the bees, are important in the life cycles of many plants as pollinators. Third, those that are part of the food web, like crickets, which feed the lizards. He added another criteria as well: aesthetic value. Biosphere 2 would have butterflies.

Inside Biosphere 2, however, there'd be no spraying for insects, since they provide indispensable services within an ecosystem. Termites, for example, are capable of digesting the cellulose of dead wood, passing on waste products that other insects, plants and some animals can use. Termites are such voracious eaters that SBV conducted "termite taste tests" to assure itself that the species chosen wouldn't eat the sealant used to give the spaceframe its airtight closure.

By June 1990, SBV's insect list was nearing finalization. Silke Schneider, who oversaw the quarantine insectary, was busy making collections and applying for permits to assemble some forty-two species consisting of two hundred and fifty colonies in addition to more than twenty thousand individual, non-social insects, which would be released into Biosphere 2. These consisted of first-priority insects, such as detritivores, predators, and pollinators crucial to any ecosystem, and second-priority insects, satisfying aesthetic needs.

NEIL RETTIG

Termites are necessary to ensure the recycling of dead plants.

The SBV Insectary office. Insects have evolved the greatest number of species of all animals in Biosphere 1 and play key roles in Biosphere 2.

KENT WOOD

C. ALLAN MORGAN

Above: Silke Schneider making observations in the Insectary.

Right: Ant eating a glue-spitting termite.

Below right: Ants, among many of the insects needed, are bred in the Insectary for introduction into Biosphere 2.

MARGARET COLLINS

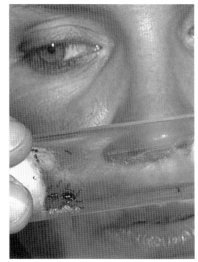

C. ALLAN MORGAN

The rainforest was getting a squadron of detritivores from Puerto Rico, including two colonies of termites, some twenty-four colonies of ants, five thousand millipedes and one thousand specimens of *Blaberus*, a genus of cockroach. That much-maligned insect had found appreciation among those designing Biosphere 2, who understood the important role insects perform in decomposing organic matter.

Other pollinators included the white line sphinx moth, an Arizona variety that rivaled a sparrow in size, who would live in the wilderness area. A variety of bees, including a colony of honeybees would perform the crucial task of pollinating the crops in the agricultural area. "That does not necessarily mean the biospherians will eat honey," Silke Schneider, manager of the insect production at SBV said, "because the bees may well need all of that for themselves." Bee "pastures" of flowering plants will grow on the central pillars in the farm area of Biosphere

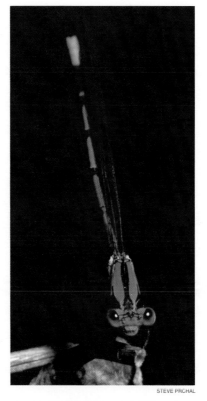

STEVE BUCHMAN

STEVE PRCHAL

STEVE PRCHAL

Yet more citizens of Biosphere 2: Carpenter bees and dragonfly.

2 to ensure enough pollen and nectar to support the honeybees.

The upper savannah was to get sixty colonies of ants, many of which spent months before closure living in terrariums with masked test tubes as home for pregnant queen ants. Schneider expected the move to Biosphere 2 could mean as little as moving each of the test tubes, carrying the queen and a contingent of worker ants. The task would be different for the ten colonies of termites slated for the lower savannah; although some termites nested in logs, others lived in fragile nests constructed from mud and their saliva, and needed some care in moving.

The desert was to get skunk beetles, ground crickets, camel crickets, sunroaches and the common fruitfly, *Drosophila*. Bumble bees and leafcutters also would pollinate desert plants. On the aesthetic list were the zebra butterfly, South American varieties but which are now raised in the United States by commercial insectaries. Many of the insect species could be collected in the United States, including the grounds of SunSpace Ranch, where Schneider and compatriots would set out blacklights to attract insects for capture at night.

The Bet

W hat will be the level of extinction? Some species are bound to die out, but in what proportion? Leigh and noted ecologist Howard T. Odum have a bet. Leigh, believing that proper human management of the biomes will prevent imbalances in the interconnected ecosystems, predicts that fewer than twenty percent of the thirty-eight hundred species will succumb. Odum bets that only twenty percent will *survive*.

5
Technics

Doom! Doom!
You have destroyed
A beautiful world
With relentless hand,
Hurled it in ruins,
A demigod in despair!
We carry its scattered fragments
Into the void
Mourning
Beauty smashed beyond repair.
Magician,
Mightiest of men,
Raise your world
More splendid than before,
From your heart's blood
Build it up again!
Create a new cycle
For the splendors of sense to adorn;
You'll hear life
Chant a new and fresher song.

Goethe

A New Technology

Lewis Mumford had predicted that the next phase in technological evolution would need to be *biotechnics* — a technology based on and compatible with life systems. Others have come to similar conclusions regarding, if not the inevitability, then certainly the necessity for such a shift. John Todd, for example, one of the creators of the New Alchemy Institute, looks toward the synthesis of architecture and life sciences in *Sustainable Communities:*

"On a practical level, ecology can be the basis for a design science ... it provides the framework for fascinating linkages in science and technology that can be applied to human settlements. Polymer physicists begin to invent materials that function like our skin, or like the terrestrial atmosphere. Electronic, computer and information networks couple to food-producing ecosystems and provide a memory and story of biological phenomena that in turn influence design and architecture. Buildings become 'organisms.' Villages and cities can be developed within this context."

Mumford, a keen student of history had seen — long before the communications satellites were launched and the revolution in Silicon Valley liberated the computers — that the city was part of the scheme of nature:

"Organisms, societies, human persons, not least cities are delicate devices for regulating energy and putting it to the service of life. The chief function of the city is to convert power into form, energy into culture, dead matter into the living symbols of art, biological reproduction into social creativity. The positive functions of the city cannot be performed

GILL C. KENNY

Signalling the giant crane operator for the placing of the last struts tying in the space frame for the west "lung" (variable expansion unit) of Biosphere 2.

Opposite: Sealing on of the specially designed sandwich glass panes.

59

A view from inside the west "lung" just before glazing the top of the ceiling.

without creating new institutional arrangements, capable of coping with the vast energies modern man now commands: arrangements just as bold as those that originally transformed the outgrown village and its stronghold into the nucleated, highly organized city."

The imperative was certainly clear to the Biosphere 2 design team. If it were to contain contemporary man, Biosphere 2, like Earth, *needed* a technosphere. Tons of matter would have to be transported, organized into structural, mechanical, electrical and electronic infrastructure to create a home for life analogous to the many geological functions of the planet. Because of the smaller scale of Biosphere 2, its technosphere needed also to stand in for the vast weather and climate systems of Biosphere 1. And, as an experiment and a work in progress, the environment of Biosphere 2 had to be electronically monitored to an unprecedented degree. Unlike the present technosphere of Biosphere 1, the technosphere of Biosphere 2 had to be coordinated, in some cases subordinated, to the needs of life.

Historically, the technosphere of the Earth had been largely designed for purely local profit and power objectives, with little regard for its effects on ecosystems. It is an increasingly contentious subject as environmental lobbies demand that technical systems not destroy their surrounding ecology. A wry comment from one scientist described well this old-style adversarial relationship between the technosphere and biosphere: "Humankind has waged war on the biosphere, and I am sorry to report that we are winning."

In Biosphere 2, mechanical devices would have to perform some natural functions. The glazed spaceframe would hold in Biosphere 2's atmosphere just as the Earth's gravitational pull keeps the planet's atmosphere from dissipating into space. Pumps circulate water and fans blow air, but only as a back-up system to the Sun. A wave machine would keep water moving around the reef in the ocean, an ecosystem whose existence depends on the rhythmic pulsating of water.

Mountains, stream beds, waterfalls, caves, differences in elevations and formation of rocks, a role normally played by the geology of Earth, were to be the work of the architects and construction team. Bulldozers would sculpt the ocean floor; men and women would build the coral reef. The power of the technosphere was harnessed for the creation of natural splendor, quite a contrast to its usual application.

The design, construction, and operation of an air-tight, three-acre, enclosed ecological system able to provide its own internal weather and recycling systems was a monumental engineering and architectural achievement. And perhaps no part of Biosphere 2 would be as revolutionary as the technosphere designed in order to serve the life systems in the biosphere. Such technological advances made in Biosphere 2 could be of assistance in refining the technosphere of Earth to better serve its host.

Fan blower to help control the monsoonal wind regime of Biosphere 2.

PHIL HAWES

C. ALLAN MORGAN

Above: Mountains, stream beds, waterfalls: a variety of forms to create sufficient econiches for the unprecedentally concentrated richness of life in Biosphere 2

Left: Stream pumps to ensure the recycling of water.

JEFF TOPPING

Making an Enclosure

The spaceframe design was the brainchild of Peter Pearce, who formed Pearce Structures in 1982 and developed the trademark Multihinge Connection System upon which Biosphere 2's design is based. His early career was in design and design education; he was a staff designer in the office of Charles Eames, where he worked extensively in furniture systems design. Eames developed the tandem seating used in airport terminals around the world and the famous Eames chair. Among other influences on Pearce were Konrad Wachsmann, who designed Albert Einstein's house and under whom Pearce worked at the University of Southern California; and D'Arcy Wentworth Thompson, who linked biology and design in his landmark book *Growth and Form*. Pearce published his own views on architectural inspiration found in nature in *Structure in Nature Is a Strategy for Design*. In the late 1960s, he also served as an earlier editor of Fuller's book, *Synergetics*, developing and producing most of its illustrations. When Augustine, Hawes, Dempster, and Allen met Pearce, it seemed an intellectual-historical inevitability that they would work together.

Peter Pearce and Phil Hawes working on the design of the "microcity", the human habitat of Biosphere 2.

ROBERT HAHN

Fuller's influence is evident in the modular nature of the spaceframe design in Biosphere 2, but there is more there. Pearce felt there was something oddly incomplete about Fuller's use of the geodesic dome, and went on to put his own imprint on spaceframe design.

"There's kind of a nineteenth-century aspect to Bucky's sensibilities," Pearce said. "He was focused on this singular-key-to-the-universe idea and I think the dome was a manifestation of that. Whereas I have been known to say that the key to the universe is a combination lock."

In Biosphere 2, Pearce's spaceframe was essentially made from combinations of single six-foot tubes of steel, coated with aluminum then sprayed with state-of-the-art powder coating paint and baked to produce a virtually unscratchable white, porcelain-like finish. This finish was their insurance against corrosion from potentially high carbon dioxide levels, other potential extremes of Biosphere 2, or simple rust. Each end of the spaceframe tubes tapered into three fins, resembling the back end of a dart. A hole in each fin allowed the tubes to be joined with each other with bolts to build a multiple pyramidic geometry, forming long expanses of superstructure that were capable of forming planes, curves, or spheres. Workers would assemble sections of the spaceframe on the ground, lift them using cranes, and attach them to each other with bolts to form the upper structure resting on the concrete foundation.

Biosphere 2's structure was designed to last a century. The two-year

A view of the magnificently arching roofs of the agricultural biome, designed to catch the maximum sunlight.

experiment would be followed by at least ninety-eight one year changes of shift. For some trees, that meant decades of growth. Even after a century, Biosphere 2 would probably just be reaching "full action," as Allen calls it. The biologists emphasized that the longer the Biosphere's lifetime the more they could learn about genetics and ecology. The key question then seemed to be not how long nature inside the Biosphere would last — if it were truly self-sustaining it could go on indefinitely — but rather how long the materials out of which the Biosphere was constructed would last.

Walter Adey pointed out that trace elements and other potentially dangerous metals would leach out of many types of stainless steel exposed to sea water. It would do no good to make the bottom of the Biosphere's ocean and marsh out of stainless steel that was impermeable to air, but which poisoned the water and the marine life.

In response, John Allen contacted Bob Walsh, a metallurgical engineer with whom he had worked back in the 1950s developing special high

Bob Walsh (right) of Allegheny-Ludlum consulting with Bill Dempster (left) and John Allen (center) on the special stainless steel liner that prevents loss of air from the below-surface levels.

ROBERT HAHN

Pittsburgh-Des Moines flooding the floor of the technical area below the agricultural biome to check for any leaks in the specially welded stainless steel.

property metals at Allegheny-Ludlum. Walsh, now a senior vice-president of the firm, agreed that ordinary stainless steel would not work well. But a special form of stainless steel, known as 6XN and produced by Allegheny, was comparable to titanium in resistance to corrosion and a predecessor to it had resisted corrosion in sea water during twenty-year exposure periods. The decision was made; SBV placed an order for about five hundred tons of 6XN. When installing it, SBV also hedged its bets by sealing the stainless steel on the inside with a layer of epoxy and further protected the epoxy with concrete. A typical portion of the foundation would be a foot of concrete, a layer of 6XN, epoxy, then another four to twelve inches of concrete.

Biosphere 2's upper regions had to permit sunlight to enter, but no air or water. Through the spring and summer of 1985, Augustine, Dempster, and others carried out studies of glass versus plastic as the covering for the spaceframe. Transparent plastic would be much cheaper than glass. It would also allow more wavelengths of light to pass through into the Biosphere. Most kinds of glass tend to screen out ultraviolet wavelengths of light. The kinds of glass that let ultraviolet in, such as quartz glass, are incredibly expensive — nearly five hundred dollars a square yard.

The light wavelength issue may affect growth of plants and the health of animals and other life forms. It was known, for example, that bees can see ultraviolet light and need it to successfully spot and pollinate some plants. It was not known whether they could manage without it. Even though transparent plastic would let in more ultraviolet light, it had one overwhelming drawback: it was simply too permeable to air.

By the autumn of 1985, SBV had decided to roof Biosphere 2 with glass *and* plastic. Dempster settled on a heat-strengthened laminated design of two pieces of quarter inch glass with a quarter inch PVB interior. The lamination through heat treatment produces a glazing that is tougher than windshield glass. Ice balls fired by a pitching machine at approximately twice the velocity of hail would not break it. Something more powerful might, but the lamination would keep the pane together with a high likelihood that it wouldn't leak.

Meanwhile, as Dempster worked on the glass versus plastic issue, he had to solve another engineering problem. Biosphere 2 would heat up during the day like a greenhouse and cool during the night and during cold, cloudy days. The expansion and contraction would subject the rigid steeland-glass structure to enormous pressures as the air inside expanded and contracted. On a hot day, pressure would push out. On a really cold day, Biosphere 2 might implode. At the very least, temperature changes might easily pop glass panels out of the spaceframe.

D. P. SNYDER

By April Dempster had reached a solution. It was an inspiration. Instead of making the spaceframe rigid enough to withstand the air pressure changes, they could make a part of Biosphere 2 flexible enough to expand and contract with the pressure fluctuations. What they needed was something like a bellows, a *lung!*

Left: Rainbow over the west Biosphere 2 lung.

Below: The air passage that goes from below the agicultural biome to the west lung. Biospherians can access the lung through this tunnel to check the structure, or make use of the large lung space as a running track.

In time, Biosphere 2 got a pair of lungs, or "variable expansion chambers". The two lungs take the form of graceful hemispheric white domes which protect the liner from the wear and tear of the Sun, but inside looks like two expanding and contracting storage tanks. One lies to the south and one to the west, each about one hundred and fifty feet away from Biosphere 2. Underground air tunnels connect them to the main structure. Inside each tank and connected to each tunnel, a gargantuan synthetic rubber membrane with a circular metal top will move freely up and down on a cushion of air. As air inside Biosphere 2 expands from the Sun's heat, it will flow through the tunnels and into the lung, raising

GILL C. KENNY

65

Looking down from the catwalk in the west lung to the variable expansion liner that keeps the inside and the outside air pressure equilibrated.

its top. As the air cools, the lungs will deflate. Instead of exerting immense pressure on the spaceframe and popping out glass panels or bursting seams, the air in Biosphere 2 will simply move in and out of the lungs.

Next, Biosphere 2 needed some kind of cooling system. Otherwise, temperatures would soar. In the basement, giant fans will circulate air through metal radiator-like grills chilled by cold water. Those fans will demand electricity, and, in a future colony on Mars, vast arrays of photovoltaic solar cells could produce that. The expense of a large solarbased power system for Biosphere 2 ruled out that approach, although improvements in technology for power storage and decreases in the cost of solar cells may lead to such a system being installed someday. In the meantime, large, natural gas-powered generators will provide Biosphere 2's electricity as part of an energy system that also will pump hot and cold water through a system of closed, heat-exchange piping in Biosphere 2 for heating and cooling.

Step by step, all of these solutions were getting the creators of Biosphere 2 closer to their goal. But the engineering problem weighing most heavily

GILL C. KENNY

These cooling towers are huge artificial waterfalls that can be used to control the temperature of Biosphere 2 on the hottest summer days, without breaking the integrity of the sealed structure.

on the mind of project director and co-architect Margret Augustine was making sure that the roof didn't leak. Nobody had ever built an airtight building the size of Biosphere 2 from glass and steel. It was generally regarded in the construction industry as impossible. SBV wanted to shoot for a leakage of one percent each year. That standard meant that over the course of a century the air inside Biosphere 2 would change-over only once.

Finding any leaks once the glazing was in place would take some doing. A *single* hole, a tenth of an inch in diameter, would leak the entire one percent amount of air allowed for. That would mean hunting it down along Biosphere 2's fifty miles of glazing seams plus another twelve miles of welded seams, most in the stainless steel liner. In one respect that would be child's play. The greatest concern wasn't the fear of one hole a tenth of an inch across. It was one hundred holes each one one-hundredth of an inch across or ten thousand holes one one-thousandth of an inch across!

In about the middle of 1985, SBV had started contacting companies that specialized in glazing to exchange ideas and develop approaches to the problem of sealing Biosphere 2's glass roof. The space frame was metal, impermeable to air just like the glass that would be attached to it. The problem was the seal between them, which required microscopic perfection. A minute piece of dirt or dust would thwart the best of seals between a hard glass and a hard metal surface. Wind also would flex the spaceframe. So, the seal between the metal and glass had to endure the push and pull of movement by its neighbors and still prevent air from passing through.

Early attempts focused on non-setting compounds with a consistency similar to bubble-gum. Dempster called them "goo." Because they didn't cure like glue, they'd remain moist and soft, and presumably flex with the expansions of the glass and metal. To test the approach, SBV hired a San Francisco firm to design such a system and glaze the Test Module with the "goo." One summer day in 1986, the job was finished. Dempster and MacCallum closed the door, latched its seal, and inflated the Test Module's small lungs.

© 1991 PETER MENZEL

"We could hear air hissing," Dempster said. The Test Module leaked air
so badly that it was audible. That meant they were nowhere close to the
one percent goal. Worse, they could not locate the holes.

The experience convinced Dempster that talking to most glazing
contractors about an airtight building was like speaking a language
they could not understand. Early in 1987, Pearce came forward with his
own proposal for putting the glass directly onto the space frame. Pearce
was able to assemble and test it, under Dempster's critical eye, convincing
SBV of its merit.

Settling on a design was only half the problem of reaching the one
percent leakage standard. No one expected installation of two hundred
and two thousand square feet of glass and metal panels to occur
perfectly. Leaks were a certainty. What SBV needed was a method of
readily finding and fixing them. The challenge seemed overwhelming.

But any kind of leak test couldn't begin before the spaceframe and
glazing were finished. The agricultural area could be closed off from the
wilderness for separate testing, but even that would not be possible
before July 1990. That pushed the entire testing schedule into the fall.
Dempster was beginning to concede that a full testing for leaks could
not be done before the period of partial closures.

There was one reason for optimism. When the Pearce spaceframe and
glazing were put atop the Test Module, SBV had been able to detect and
repair leaks in sealings around apertures — the door, electrical conduits
and welded joints. The leakage rate amounted to about eight percent a
year, meaning the air inside would change once every twelve years.
Because Biosphere 2 was much larger — a smaller ratio of spaceframe
area to volume — any engineer would expect better performance.
Dempster calculated the equivalent leakage for Biosphere 2 just over
three percent a year, meaning a cycling every thirty years. It was hoped
improved quality control in application might boost that close to the
now almost legendary one percent goal.

Managing the Water Cycle

Earth's biosphere depends on a natural water cycle. Energy from the sun evaporates water from lakes, oceans, and other bodies of water. Water vapor builds up in the atmosphere, clouds form and become saturated, rain falls, which feeds streams and rivers that eventually return the water to lakes and oceans.

Since Biosphere 2 was too small to ensure the complete water cycle occurring naturally, its designers would have to figure out a way to *make* it occur with some mechanical assistance. The task was made more complicated by the demands of controlling levels of acidity and alkalinity; acidic water flowing from the rainforest would affect the neighboring savannah, whose plant life depends on less acidic water. Fresh water also would flow into the ocean, making necessary a method for recovering fresh water again — or else all of the fresh water in the wilderness would eventually mix with the sea water, leaving a brackish brew. SBV settled on a design it hoped would wed mechanical systems with natural forces, supplementing the natural evaporation from the ocean with a system to desalinate the amount of water needed to maintain the fresh water reservoir.

The closest thing to rainfall expected in Biosphere 2 was an artificial cloud in the top of the rainforest biome generated by misting machines, a necessity to maintain the health of rainforest species accustomed to perpetual high humidity, and condensation on the glazing that would drip. The drips and fog will modestly feed a pool atop the rainforest mountain, whose major inflow will come from water pumped out of reservoirs in the basement below. The pool feeds the rainforest stream, which meanders toward the savannah. Before reaching the savannah, however, most of the water in the rainforest stream will be diverted to pipes leading it back to the basement reservoirs, recycling it within the rainforest.

The mist rises in the Biosphere 2 cloud forest.

An average of about three to three and a half gallons a minute will cascade into the savannah stream, which also would have its own system of recycling water. About three gallons a minute will flow onward to the freshwater pond, and from there about three gallons a minute will feed the fresh water marsh, which in turn will feed the salt marsh. On average, as the tide of the salt water marsh rises and falls, the same rate of flow will feed into the ocean, completing a mountain-to-the-sea journey in a matter of a few hundred feet. Desalinization equipment will extract about three to ten gallons a minute, depending upon the season, from the sea, remove its salt for return to the ocean, and send on fresh water to the basement reservoirs below the rainforest, completing the cycle. This didn't sound too bad. Keep the pumps and the desalinization going, and the sun and gravity would do the rest.

Sprinklers can be set to back-up any lacks in the rainfall created by the interaction of rainforest and architecture.

GILL C. KENNY

But other additions and subtractions from the principal flow of water complicates management of the water flow, and requires a system for making adjustments and keeping track of the location of water in Biosphere 2. The net flow of three to three and a half gallons will require adjustment to compensate for evaporation along the way and for absorption by plants growing in the marsh or along stream banks. Along the sea cliffs, permanent misters will irrigate the cliffside plants and make small donations to the ocean. Water will evaporate from the ocean, condense on the glass, and flow down it into a gutter system that will feed small reservoirs and eventually feed into the principal stream system. "Rainfall" will also come from drip and sprinkler irrigation systems or manual waterings by the biospherians. And water percolating down through the soil will reach a grid of drainage pipes that will return it upstream again.

The major component of plant tissue is water. As the biomes of Biosphere 2 grow, the increase in living tissue will tie up larger and larger quantities of water. Without additional water, the growth of plants would slowly reduce the amount of free water. To compensate, a two hundred thousand gallon reservoir housed within one of the two lungs will act as a source and buffer, feeding into Biosphere 2 as needed.

The agricultural biome will have a similar reservoir in its basement, which will feed irrigation water to plots growing potatoes, beans, greens, squash, peas, and other grains, vegetables, and fruits. Here, the cycle will be simple. Irrigation water percolating through the soil will return to the reservoir for another cycling as irrigation water. Water condensed in basement air coolers, which also function as dehumidifiers, will replace water lost to evaporation or plant transpiration. Or, if necessary, the biospherians will move water between the wilderness areas and the agricultural biome to compensate for a lack or surplus of water.

Most difficult of all, wastewater produced by the humans had to be integrated into the water cycle. Bill Wolverton already had an answer for this problem: use plants and microbes. His work with plant-based water treatment systems at NASA Stennis Space Center in Mississippi showed that plants can and do absorb more than just carbon dioxide. They can be used as living air and water filters, which also produce beneficial oxygen and plant mass. His system had already been tried out successfully in the Biosphere 2 Test Module.

Wolverton helped pioneer similar uses of plants and associated microbes to clean water, mimicking the characteristic ability of marshes, swamps and other slow-moving, vegetation- and microbe-rich wetlands to take in raw sewage and produce clean water. Because creating wetlands is relatively inexpensive, undeveloped countries and municipalities with low tax bases have used the method to treat wastewater and clean up industrial pollutants.

Calculations showed that a small tank inside Biosphere 2 could handle the volume of wastewater that Biosphere 2 would produce. But the challenge of using such a system within a closed system was new. Not only would its effects on the entire closed ecosystem have to be assessed, but it would have to be made more efficient and miniaturized so that it wouldn't take up a disproportionate area as well.

Ecotechnics

Technology is more than a matter of machines. It is fundamentally the logic of systems designed to fulfill purposes. Bringing technology into the service of life has been one of the most revolutionary aspects of the making of Biosphere 2. By putting the organic systems first — in all their awkward complexity and splendid diversity — the technology required has turned out to be of the most up-to-date and sophisticated, fulfilling a prediction made many years ago by Lewis Mumford. Every machine has to be energy efficient and also made fail safe. Every working part has to be modest in emissions. Above all, in the very center of the most sophisticated system of monitoring and control, room has to be made for the key representative of life on Earth in this project — humankind in its two-fold role as keystone predator and technologist.

Life, the biosphere, creates its own controls and balances. From studying this our insight into world processes is conceived. This is something that all involved in the project have had to learn over and over again, more deeply each time, and always left asking, "is there anything more?"

6
The Farm

Oh Adam was a gardener, and God who made him sees
That half a gardener's work is done upon his knees
So when your work is finished, you can wash your hands and pray
For the glory of the Garden, that it may not pass away!
And the glory of the Garden, it shall never pass away!

Rudyard Kipling

Intensive Agriculture

Carl Hodges of the Environmental Research Lab (ERL) at the University of Arizona in Tucson, initially began agricultural research for Biosphere 2 using sand as the growth medium. This type of agriculture, often inaccurately described as soil-less agriculture, uses sand merely as an anchor for plants, whose roots are fed a simple solution containing only the basic needs for their growth. This method, as far as ERL had determined, maximizes crop yields in very small areas. On the Moon or Mars, with space at a premium, the ability to produce great amounts of food in small plots would be desirable.

However, he came to realize that maximizing the yield of Biosphere 2's agriculture wasn't the only issue: the overriding concern was *keeping the cycles going*. The biospherians might eat well off of small plots with sand culture, but if the entire ecological system inside collapsed, they would have accomplished very little.

A lot of time and money was spent exploring artificially-fed crop growing — not only hydroponics, where nutrients are fed in solution, but also aeroponics, where roots are not even immersed in liquid but hang in the air and are sprayed periodically. The Institute of Ecotechnics team could not overcome the prevailing assumption in technical circles that life had to be strictly controlled, and this meant that microbes were to be eliminated by sterilization. Ultimately, this "futuristic" notion proved to be incompatible with the recycling criteria of Biosphere 2.

Sand culture ignored an important role of the soil; that it is the route through which dead plants and animals are recycled, ultimately, and that it can play a role in cleaning the air. Bios-3 had no soils and as a result it suffered from problems in air quality. Bacteria in the soil are important in cycles of carbon, nitrogen, sulfur and other elements that are the building blocks of plant and animal tissues. Along with detritivores — waste-eaters — such as soil nematodes and other creatures that live off dead plants and animals, bacteria help recycle nutrients. Without them, dead animals and plants would simply pile up until Biosphere 2 became a large heap. Some bacteria also help fix nitrogen so that plants can absorb it.

Nevertheless, all this preliminary work proved extremely valuable. The Environmental Research Laboratory's careful studies under controlled conditions proved beyond a doubt that the biosphere "knew" best. That and the stench from Shepelev's chamber made clear to all the fact that Biosphere 2's agriculture, as well as the other biomes, needed soil because it was inextricably linked to the atmosphere, the life cycles.

© 1991 PETER MENZEL

Carl Hodges, Director of the Environmental Research Lab of the University of Arizona.

RAY MANLEY STUDIOS

Some of the first intensive agriculture that led to the selection of the more than 140 crop species chosen from 1500 studied.

Opposite: Two biospherians, Jane Poynter and Linda Leigh, harvesting an experimental plot of sorghum.

C. ALLAN MORGAN

*Examining the soil for moisture
and tilth.*

Soil

In his book *Farmers of Forty Centuries*, published almost ninety years
ago, F.H. King wrote:

*"While it was not until 1877 to 1879 that men of science came to know
that the processes of nitrification, so essential to agriculture, are due to
germ life [microbes], in simple justice to the plain farmers of the world,
to those who through the ages from Adam down, lived close to Nature and
working through her, have fed the world, it should be recognized that
there have been those among them who have grasped such essential, vital
truths and kept them alive in the practices of their day."*

The important step in developing Biosphere 2's agricultural biome was
the decision to use a compost-based soil. It was realized that the project
goal of a sustainable system tested and observed for one hundred years
would require the continual recycling of animal and human wastes,
and inedible plant residues via earthworm farms and compost piles.
Compost is a method of accelerating the natural soil-building
mechanisms which normally require a century to make one inch of new
topsoil. In the compost heap the mixture of manure, vegetable materials
and a bit of soil to ensure the presence of soil microbes heats up,
destroys possible pathogens and starts the decomposition process.
After the compost cools down, earthworms are added to assist its
transformation. In one to six months, depending on how the compost
is handled, a sweet-smelling black topsoil is produced.

The testing of prospective crops for Biosphere 2 occurred both at the
SBV's Biospheric Research and Development Center of greenhouses
and at greenhouses at ERL. Production for many crops exceeded
expectations — especially bountiful were sweet potatoes, peanuts,
cowpeas, leuceana, and elephant grass. An early trial at ERL with the
consulting of John Niederhauser, a famous potato agronomist, produced
a world record crop in the two foot square stacked box method of
production he designed specially for the project.

By happy coincidence, the first movement of the soil into Biosphere 2's
intensive agriculture biome occurred on July 21, 1989 — the twentieth
anniversary of the Apollo 11 landing on the moon. A large gathering of
SBV staff, consultants and construction personnel gathered to help
shovel the rich soil mix into the waiting wheelbarrows. So attuned had
people gotten to the concept of biospherics that it became clear why this
was a signal moment — the soil, rich in microbes, was the first life to
enter Biosphere 2. Even more astonishing, according to Carl Hodges,
was hearing this coming from the president of one of the large
construction companies working on Biosphere 2. "When a construction
engineering company executive announces that it's a red-letter day
because the first life — the first soil — is going into Biosphere 2, I knew
we had made some real headway in terms of biological awareness."

"Soil is your basic wealth," biospherian and farm manager Jane Poynter
says. "If you manage your soil well it'll work." But finding the right soil
and getting it into Biosphere 2 was not easy. The job of finding it fell to
Stephen Storm, Director of SBV's Tissue Culture Laboratory and a long-
time practitioner of composting.

MARIE ALLEN

The first planting of the intensive agricultural biome, the most productive agricultural system yet developed - and it's long term sustainable with no toxics applied.

"In the beginning, when all things seemed possible, we talked about bringing soils in from all the biomes of the world," said Storm in reference primarily to soils for the wilderness biomes. "Then, as reality intruded more and more upon our dreams we began to concentrate on soils from the United States — ancient rainforest soils beneath Georgia — that sort of thing. But we found out that not only are soils virtually prohibited from entering the United States unless they are sterilized — thus ruining their viability — but Arizona as well has tough regulations on soil importation. Transportation costs were also prohibitive, so we started to look right around here."

It wasn't until Storm was told of Wilson's Pond that they found what they were looking for — just five miles away! Throughout the years Wilson's Pond had been used as a stock tank. When it filled with water, cattle came in to drink, fertilizing the soil around the pond with dung. As the water receded, shrinking the pond to a smaller and smaller circumference, the cows worked their way toward the center, trampling dung and other organic materials into the soil. Each year the summer rains would wash in desiccated cactus pads, twigs, leaves, mesquite beans, sticks, and more cow dung. The pond was heavily silted in with rich alluvial soils and a fine crop of composted manure.

Using material dredged from the pond as a matrix, the agriculture group began to build their soil. The mix they finally came up with is about seventy percent alluvial soils from Wilson's Pond, about fifteen percent peat, and fifteen percent compost. When they finished they had a soil with a neutral pH factor of around 6.8, meaning it was neither too acidic nor too alkaline, a good medium for growing almost anything.

Opposite: The core of the intensive agricultural unit consists of 18 fields which are rotated on a calculated basis.

Biospheric Farming

Every farmer knows that you can't keep taking things out of the soil, no matter how rich its original condition, without putting something back. Soils have to be replenished. Conventional farmers do this in three ways: they can let a field lie fallow for a season from time to time. Or they can compost animal and plant wastes and spread the mix on the land during the winter when no crops are under cultivation. They can also fertilize crops directly by either adding chemicals like nitrogen to the soil, or by growing 'green manure' crops that are ploughed back into the soil.

Only two of those options are available to biospheric farmers. Because the growing area of the agriculture unit is small, the ground must be under cultivation all the time. No plot can lie fallow for longer than a few days. And, in a closed system, the introduction of chemical fertilizers is not an option unless they are manufactured on the spot. Adding compost to the soil and growing 'green manure' are the only available choices.

As any gardener knows, composting takes time. Most growers like to compost wastes for six months to a year before adding them to their garden plots. But in the Biosphere 2 garden everything needs to be recycled fast, which means that even the rate of decay in the compost heap must be accelerated if possible. To help do this, a specially designed compost machine was installed in the basement below the garden.

Of course there will be earthworms in the soil, red wrigglers and one larger type called the jumper worm. Bat guano has been imported and added to the soils. Of all animal fertilizers, bat guano is the richest in diversity of its microbes which are necessary to break down organic materials in the soil.

SBV named the farm unit the *intensive agriculture biome (IAB)*. And intense it will be. The farm consists of eighteen garden plots ranging in size from just under five hundred square feet to almost nine hundred and fifty square feet. To comfortably feed eight people, biospherian Jane Poynter hopes to jam three crops a year into each plot, except for plots set aside for perennial crops and fruit trees. In conventional agricultural systems on the outside only one or two crops are planted per year in a given plot. Where she can't manage three, she will settle for two. Between crops the soil has to be worked to prepare it for another planting and allowed to dry out if it gets a bit wet, or watered if it's too dry. The manner in which crops are rotated is based on dietary needs, soil conditions, and on the necessity of maintaining soil fertility. Another consideration is what plants should follow one another in a given plot.

"Some crops simply can't stand to come in after other plants," says Poynter. "Pinto beans hate oats. Once I made the mistake of putting beans in after oats. I'd very carefully mulched the oat hay before putting in the beans. Nothing grew. I couldn't figure it out. Certain plants put out chemicals that others can't stand. Certain other things won't grow after sweet potatoes, for instance."

C. ALLAN MORGAN

The intensive agriculture is based on sub-tropical climatic conditions and is modelled to be of use to the hundreds of millions of small farmers living in these regions as well as being maximum conditions for production in space colonies.

Farmers must also be aware of what a plant will add to or take away from the soil. Peanuts, which are legumes, actually replenish the soil as they grow and seem to do better if left alone in the same rotation for a couple of years. Peas and beans are used in other crop rotations because of their ability to fix nitrogen in the soil. In deciding on what crops to put in and how to plan rotations, Poynter was guided by another key consideration for the Biosphere. What would actually grow inside? The atmosphere inside is tropical and humid, with summer highs averaging eighty-five degrees F., and winter lows averaging sixty-five degrees F. The light, however, coming into the garden area is of a temperate zone which has a shorter winter daylength than found in the tropic zone. Some crops are affected by this; sweet potatoes for example, simply grow lots of leaves and no potatoes without sufficient daylength and will have to be grown only in summer.

Enough of Everything

Although crop rotation and soil replenishment were important criteria for both crop selection and rotation, the guiding criterion was what would provide an adequate diet for the crew inside. What could be grown in a space slightly over one-half acre that would meet the dietary requirements of eight humans? The answer to this question became a project for a team of agronomists from the Environmental Research Laboratory, a nutritionist from the University of Arizona, and two from SBV. The original diet plan was vegetarian, consisting of legumes, grains, leafy greens, oil crops such as peanuts, and crops in the starch group such as white potatoes, sweet potatoes, and squashes.

All the dietary needs of humans can be met on a strictly vegetarian diet. Combinations of legumes and grains, for example, produce a complete protein. The only problem with a vegetarian diet, or even a predominantly vegetarian diet, is getting enough fat. These days with most people aware of the harm that too much fat in our diets can do, we tend to dismiss it as an unnecessary constituent of a balanced diet. But fats provide more energy per gram than other foods, and are important sources of the fat-soluble vitamins. So, for the hardworking biospherians, pigs, chickens, and goats will be part of the human diet. Even so, getting enough fat will still be a problem, since the meat available will provide only a part. Certain vegetable crops, like peanuts, are high in fat, though not high enough to provide completely for dietary requirements; but together with the animals, the necessary fat should be provided.

If all goes well the crew should have plenty of everything else to eat. There will be rice, wheat, barley, sweet potatoes, pinto beans, papayas, bananas, swiss chard, squash, tomatoes, strawberries, onions, eggplant, corn, watermelon, carrots, and several varieties of white potatoes, and, of course, eggs and goat milk, the occasional coffee, hot chocolate, and tea.

In the small tropical orchard the Biosphere crew will harvest lime, avocado, guava, fig, tangerine, lemon, papaya, banana, mandarin, kiwi, pineapple, grapes, star fruit, kumquat, and tropical apple. No space is wasted. Small areas around the edges of the garden and orchard are planted in passion fruit, sugar cane, pineapple, melons, kale, and Poynter's assortment of herbs. Mung beans will grow beside bananas, artichokes next to peanuts, papayas next to coffee trees. White potatoes, chiles, bush beans, and earthworms will even grow in shallow beds in special fiberglass boxes set atop air vents. Coffee, banana, and papaya trees will grace the north end of the sorghum plot and soybeans will separate the rice paddies. Cacao and tea will be some of the few harvested crops from the rainforest biome.

Papayas, bananas, swiss chard are some of the contributors to the rich yield.

Domestic Animals

Beneath the apartment of Bernd Zabel in the west wing of the human habitat special bays have been set aside for domesticated animals. Zabel is their keeper. Every morning he will awaken to the crowing of the roosters. He will descend from his bedroom to the warm and humid "barn" below to be greeted by the bleating of the pygmy goats and the snuffling of pygmy pigs.

For the next two years, Zabel will feed and care for the goats, the jungle fowl and the pigs. He will supervise their breeding and births. He will give them vitamins to keep them perky and care for them when they are sick. Eggs from the chickens and milk from the goats will supply about twenty per cent of needed protein.

A miniature world with a miniature farm needs miniature animals; and, because more can be kept, there is greater genetic diversity. The goats are African pygmy goats from Nigeria. The small hardy feral pigs require little or no medical attention and are extremely resourceful, being adepts at independent rooting and foraging and willing to eat the most unpromising stalks and leaves to produce manure for compost. Keen and intelligent like all pigs, the little piglets, when barely four days old, surprised their keepers by locking necks and running round each other in a sort of pig version of wrestling. The chickens include a jungle fowl, gallus gallus, the original chicken species from India, Japanese silky bantams and a Space Biospheres cross between the two. All the animals have good temperaments and are extremely heat tolerant.

The special Space Biospheres breed of chicken: a cross between the tough original jungle fowl from India and the refined Japanese silky bantam.

KENT WOOD

80

C. ALLAN MORGAN

Highly efficient and good-tempered pygmy goats from the Plateau region of Nigeria.

Four pygmy goat does and one billy will be taken into the Biosphere. Two sows and one boar pig, along with three roosters and thirty-five hens have been selected. All the domestic animals will breed by natural methods. The main function of the goats is to provide milk, so at all times, three of the four goats will be lactating, while the fourth is being bred. Goat milk will be the highest source of fat in the human diet, and pygmy goat milk contains the highest butter fat content of any goat milk.

The pigs and the goats eat what no one else will. Zabel hopes to breed the two sows twice a year to produce a total of twenty piglets on average. He hopes to add about forty-five pounds of goat meat and two hundred pounds of pork to the diet of the crew.

Silke Schneider trimming the goat's hooves.

C. ALLAN MORGAN

The reason for cross-breeding silky chickens with jungle fowl is that the crossbreeds are somewhat more docile without losing their jungle-smarts, allowing Zabel to raise a greater density of chickens. The thirty-five laying hens are expected to produce on the yearly average eleven eggs per day, eight for human consumption and three for brooding; in addition the chickens will add thirty-five pounds of meat per year to the diet for humans.

All the animals selected are gentle and easy to handle, which is important in such a small closed environment. The pygmy goats and even the feral pigs enjoy human contact naturally, and successive breedings in the Biosphere 2 research facilities have made them even friendlier. The application of ethology (the study of animal behavior) by Diana Mathewson, Director of Agricultural Systems, has helped immensely in facilitating a productive and happy animal system. The domestic animals - pigs, chickens and goats -living together in Biosphere 2 have established their own social order. To achieve this integration, their environment was designed to allow for the range of behaviors that each of the species enjoys. Friendly and content animals produce more.

The goats will subsist mainly on fodder crops like siratro, elephant grass, and leucaena, a leguminous tropical tree with leaves that provide excellent forage high in vegetable protein. These are perennial crops that can be grown summer or winter. The pigs are tropical animals used to subsisting on jungle foods. They will be eating primarily leftovers from the human diet, supplemented with starchy vegetables — malanga and yautia — and crop roughages. Worms will provide the main protein source for the chickens, and a small amount of sorghum will also supplement their diet.

Special fodder crops are grown to supplement the goats, pigs and chickens' usual diet of waste from the field crops and dinner scraps.

C. ALLAN MORGAN

The ancient fish, rice, and azolla system from China with specially bred strains of rice and fish to increase production.

The fish-rice-azolla ecological system was developed and managed by Bernd Zabel, biospherian and general manager for the construction of Biosphere 2. After about four years of trial and error and experiments in the research greenhouses Zabel has come up with a virtually self-sustaining fish-rearing system. It involves growing rice, tilapia fish, and azolla — a small floating green fern — in the same paddy. The highly productive breed of tilapia was developed by the Environmental Research Laboratory.

Although very old, having been used for thousands of years in China, the three-fold system had to be fine-tuned for the intensive agriculture practiced inside the Biosphere. For one thing, unlike the huge rice paddies in China, Biosphere 2 paddies are very small. Six of them measure only fifteen by eighteen feet each, a total of two hundred and seventy square feet; and eight smaller paddies in the basement add another thirteen hundred square feet. So, while bumper yields of rice would be produced in that space, if the crew of eight were going to eat fish, preferably a couple of times a month, Zabel has had to work to maximize the fish yields proportionally. The fish would have to grow to maturity quickly, be harvested, and a new crop started right away.

C. ALLAN MORGAN

Right: Bernd Zabel examining the set of rice from one of his paddies

Below right: The azolla, or water fern, in the fish-rice system multiplies rapidly with sunlight, providing food for the fish and nitrogen for the rice.

C. ALLAN MORGAN

Zabel began looking for ways to supplement the diet of the fish to make them grow faster. The answer was earthworms, red wrigglers. In one test plot, the fish were fed a daily handful of worms to supplement the azolla diet. They grew faster. At any given time no more than approximately fifty fish will live in this fish-rice-azolla system of the intensive agricultural biome. "At this rate, we'll have fifty pounds of fish a year," Zabel says, not to mention the bumper rice crop that grows above them. "That gives us a meal of fish once a month as a nice bonus to our rice production. I may be the first German rice farmer in the world," he laughs.

One thing that there will be plenty of according to Mathewson, is earthworms. In a pinch they can also be added to the diets of the pigs. "Worm production inside the Biosphere will be great. Every seven to ten days an earthworm can produce one egg capsule containing three to twenty fertile eggs. The eggs hatch within two to three weeks. Special boxes for worm production are placed throughout the intensive agriculture providing ample quantities of high-protein food supplement for the animals. Even pigs love worms. A lot of people don't know that — but they really do love them."

The Little Critters

Insects wandered into Biosphere 2 off the desert during construction and took up residence in the various developing biomes. Some of these species are beneficial to humans, some detrimental. The "good" ones are carnivore insects that prey on "bad" insects that munch on food crops. Impressive evidence of the presence of good-guy and bad-guy insects came early in April 1990 when a spring rain at the site was followed by unseasonably warm temperatures — prime conditions for insect explosion. And explode they did. Suddenly aphids were everywhere, gobbling up alfalfa plants. But almost as sudden as the infestation of aphids was the reaction of predator insects. It was as if "food scouts" had been sent out from insect outposts in the nearby desert. Armies of ladybugs, parasitic wasps, and brown lacewings moved in, swiftly annihilating the aphids. Oddly, both events were a blessing. Poynter knows that detrimental insects, wanted or not, will invade her gardens. She also knows that the invaders will be opposed by beneficial insects. But the beneficial ones will stay on the job only so long as there is something for them to eat. If the aphids are poisoned or die off completely for some other reason, the ladybugs will just as surely die.

One of the oldest and most labor intensive methods of insect control is to remove them by hand. Sometimes this will mean that Poynter and the others will walk the garden rows, plucking and squashing unwanted bugs. Poynter may even enlist the support of her chickens from time to time. "If I get an outbreak of something in a particular plot that I know chickens like to eat, I've got a nice little portable bamboo fence that I can set up, and put them on it for a day, let them munch, and then return them to their pens at night." But Poynter is not one to leave it at that — she has a number of other strategies up her sleeve as part of her integrated pest management program including intercropping techniques, trapping and barrier planting .

© 1991 PETER MENZEL

Sally Silverstone estimating her yield of ladybugs which help control excessive population growth of otherwise useful insects.

Breath of Life

The quality of Biosphere 2's atmosphere posed many difficult questions, some of which wouldn't be answered completely until the eight people were inside and the airtight door closed behind them and at least two yearly cycles had passed. Some might require a decade because the Biosphere 2 ecology would take that long to grow close to its climax state of ecological equilibrium.

If they had too many plants inside in relation to animals, the vegetation might drastically reduce all of the carbon dioxide. Then, when the concentration of carbon dioxide had dropped to fifty to one hundred parts per million, many plants — except for some survivalist species that evolved genetic capacities to process lower levels — would stop growing. So comparatively few animals could live there.

The other extreme scenario would be the result of too many animals. When carbon dioxide levels increase, plants respond to that richer diet by photosynthesizing at an increased rate. But the response is slow and deliberate, like the tortoise. In small closed areas animals could deplete all of the oxygen and die before plants had a chance to replenish the supply. Long before that, however, humans might start suffering the effects of too much carbon dioxide in the air, which may become unhealthy for humans at levels of ten to twenty thousand parts per million.

The buildup of trace gases could, like sewer gas, kill you. Even in non-lethal doses, carbon monoxide, nitrogen oxides, and other such gases affect health; methane is highly flammable. A structure the size of Biosphere 2 will have numerous sources of trace gases. Plastics, paints, sealers, and other materials used to build the structure of Biosphere 2 will slowly release various hydrocarbons. Hot lubricants will produce carbon monoxide. Electrical sparks will generate nitrogen oxides and ozone. Livestock produces methane and other malodorous gases. Plants will produce ethylene, a hormone affecting flowering, fruit maturation, senescence, and wound response. If ethylene were not cleaned from the air, it would prematurely stimulate plants into plant cycle changes. Composting of cabbage, turnips, kale, and other crucifer vegetables will also produce sulfur dioxide. This might give rise to actual acid rain!

Some trace gases, such as methane, break down naturally, high in Earth's atmosphere, but SBV didn't expect anything like that to occur in the spaceframe rafters of Biosphere 2. One did hope to take advantage of natural air-cleaning process — the slow circulation of air through the soil, where microbes consume and digest gases. Left to the natural pace, that would be painfully slow — far too slow to handle the levels of trace gases that could potentially occur in Biosphere 2. In Europe, however, scientists for years had been accelerating the process by pumping air through soil. Keeping the soil properly watered would be very important. Irrigation requires a lot of observation, getting to know a soil, how it responds to water, how much it holds and how much it flushes through.

The Institute of Ecotechnics, Lynn Margulis, and Clair Folsome had all seen the power of microbes in soil to affect the air. Hodges and his crew at ERL contracted to work on soil bed reactors and the results showed they would be very useful for Biosphere 2. The principle is simple: soil, properly aerated, contains vast surface area, all covered by a living carpet of bacteria and fungi. By pumping air through the soil beds microbes would consume trace gases quicker, digest them, and release carbon dioxide and water. In Biosphere 2, this could be accomplished by pulling air with fans and blowing it through ducts that led under a bed of soil, blowing the air up through it. By using the soil in the agricultural area — that is, giving it a dual purpose — Biosphere 2 wouldn't have to be increased in size. The air would pass up through the soil, rejoining the Biosphere 2 atmosphere among the potatoes, beans, greens, and cucumbers. As the air passed through the soil, the microbes' insatiable appetites would essentially ensure clean air. So two major obstacles appeared solvable. When SBV installed the reactor in the Test Module the eighty-five measured toxic components of the atmosphere all fell to minute levels.

Intelligent Food

Soil, crops, animals and air form a system of surprising inter-connectedness and flexibility. In Biosphere 2, human intelligence will be monitoring this system, cooperating with the natural cybernetic controls to maintain stability and attain the required productivity of high grade energy, or "food fit for humans". Life can balance its environment on many levels — even the "green slime everywhere" level that Tony Burgess joked about, is a state of balance. However, the goal in Biosphere 2 is the highest level of balance, where technological energies join with the biological ones to produce the critical energies of food, beauty and discovery.

Left: Completion of the first planting in agricultural biome.

Below: A view from the space frames of the oncoming crops.

With arable land throughout the world being increasingly lost to desertification, drought, erosion, and human mismanagement, the Intensive Agriculture Biome of Biosphere 2 may have much to tell us about how we feed future generations. The lessons learned in soil management alone may have far-flung ramifications. The aim of these biospheric farmers is to maintain high soil fertility for a period of one hundred years, the anticipated lifespan of the Biosphere. In contrast, farming with heavy machinery, pesticides, and annual applications of chemical fertilizers with huge expenditures of energy resources means that soil fertility is sometimes lost in less than a generation. The bio-

spherians' half acre is a model of sustainable, low-cost, non-polluting, high-yield agriculture. For sustaining not only human life, but an abundant world of life on a planet, its applications may be profound.

7
Architecture

"A building is of the site, not on it ... We shall take the first and most urgent idea and develop it into the final forms. In this way we are apt to come up with a truly organic performance. In nature the forms are the result of a growth and development, they are arrived at rather than started with."

Bruce Goff

The Vision and Conception

Once established at Sunspace Ranch, the architects and designers had to find the best location for the new Biosphere. Margret Augustine, Phil Hawes and John Allen walked and rode on horseback around the property, getting an intimate sense of the terrain. All their previous pilgrimages to the major architectural achievements of the world — Chartres Cathedral, the Great Pyramid, the Taj Mahal, the Dome of the Rock, Hagia Sophia and many others — and to the ancient cities of Tiahuanaco, Babylon, Ur and Konya and to modern sites of significance, such as the Kennedy Space Center and the Puerto Rican Radio Astronomy Telescope, were being distilled into the project.

Finding the right place helped the architects begin to imagine the form Biosphere 2 should take. The chosen site was protected on three sides by ridges. A great berm to be built on the north would not only provide protection, but also a panoramic ridge from which to view the finished work. On the south side, a gently sloping arroyo ran from east to west, beyond which a low ridge led on to a magnificent view of the Canyon del Oro. The Santa Catalina mountains, just above the horizon of the ridge, would be visible from inside the Biosphere, resembling a small Himalayas beyond a foothill range.

MARIE ALLEN

Margret Augustine and John Allen riding around the Great Pyramid after starting at the Step Pyramid. Part of the round-the-world site studies that started in 1975.

GILL C. KENNY

Margret Augustine and Phil Hawes contemplating possibilities of the site at Sunspace Ranch.

Opposite: Looking past saguaro cacti to the Biospheric Research and Development Center and the Biosphere 2 Panorama Ridge behind. The Energy Center is behind to the left of Biosphere 2.

CELINE/COURTESY OF MARGRET AUGUSTINE

Painting of the inner courtyard of Hotel Vajra in Katmandu, a 1978-80 architecture by Hawes and Augustine, now a famous destination point for serious subcontinental travellers.

One of Augustine's favorite projects has been the Hotel Vajra in Kathmandu, Nepal which she and Hawes designed. Capable of withstanding an 8.5 earthquake, complete with modern electricity, plumbing and solar energy, it still embodied the traditional craftsmanship and values of the people. Many people seeing it believe the building to be more than a hundred years old because it has a timeless quality. She and Hawes did not lay down a design and tell people to do it just that way: "We would do an initial design and they would come back with their ideas and there was this back and forth that really created these interesting forms."

Designing Biosphere 2 was to involve all the different specialists in a back and forth process that was often to be explosive. It was this that created something new, that none of the people would have arrived at by themselves. Unlike designing a house or a factory or an office building, they had no examples to follow in the design for a biosphere. Translating their first sketches and circles into material form was extremely difficult.

Many of the ideas that emerged in the early part of 1985 fell by the wayside as the design process continued. Hawes made a beautiful model of a sealed passageway weaving its way through the Biosphere, where visitors could walk and view the interior. Then there was the assumption that some kind of louver system would be needed to provide shade on hot summer days. This gave way to the need for maximum photosynthetic activity, leaving the energy center to provide the cooling required for Biosphere 2.

The first sketches were mere squares or circles, defining areas; but, soon some startling visions began to appear. Hawes recalls: "Margret and I locked ourselves away in the office ... it was a few days, 2,3,4 days, I don't know. That's when we actually came up with the basic shapes that we have used."

ROBERT HAHN

Left: Reed huts in the delta of the Schoti-el-Arab, near the border of Iran and Iraq, 1976.

ROBERT HAHN

Above: The great minaret of Samarra, north of Baghdad, built in the era of Scheherazade's Arabian Nights.

A major influence on these basic shapes came from a journey Margret Augustine made on an Institute of Ecotechnics' tour in 1976 to Iraq. There she experienced the architecture of many civilizations, going back to the earliest cities. What impressed her most were the reed houses still being built in much the same way as they had been thousands of years ago, made of bundles of reeds bent over then joined together, like an upside-down U. In the ancient epic of Gilgamesh, which includes the first story of the Flood, the virtuous man Utnapishtim is told by the god Ea to demolish his reed hut and build a boat in its place. Augustine recalls: "When I looked at these buildings, they actually were, if you turned them upside down, in the shape of a ship's hull." She found the same shape in later adobe buildings with their Babylonian vaulted ceilings and in the stone buildings which came after them.

Augustine wanted to use the Babylonian vaulted shape as the overhead structure of the agri-cultural biome and Hawes agreed. They began by playing around with different combinations, such as putting three in a row to form one long vault in the shape of railroad tunnel. But, Biosphere 2's tentative site was on a hillside, suggesting that one end of the biome would be lower then the other. So, they drew them into steps. Augustine reflects: "What was very interesting to me was that from the first sketch we did on paper to the first model to the final design we have now, actually in the total form or shape of things, it changed very, very little."

ROBERT HAHN

Vaults from ancient Babylon.

91

The Mission Control building houses the computer center and conference room for teleconferencing with the crew inside Biosphere 2.

The first thing that had to be actually built was the Biospheric Research and Development Center (BRDC), including space for the SARBID architectural design studios. Eight thousand square feet were needed for the tropical farm garden experiments and another eight thousand for desert, rainforest and savannah plant acquisitions to be cared for, studied, grown and experimented with. Then, there was need of an insectary, to care for the expected forty-two species which would have to be kept under quarantine conditions. An analytical laboratory and a computer center were essential. Each part of the complex, such as the analytical lab, would ultimately "clone" itself into Biosphere 2. No final move would be greater than two thousand feet.

Margret Augustine organized the site layout so that the computer center — known as "Mission Control" — would be adjacent to the Biosphere 2 site before its foundations were laid. The Mission Control building, like the BRDC offices, would take up the theme of the vaulted shapes. It would include not only the main computer complex, but also space for meetings and presentations and a veranda from which the vast construction could be surveyed as it progressed.

As for the structure of Biosphere 2 itself, nothing less than a masterpiece would be sought. There was a continuous back and forth process between the designers of each of the various biomes, the engineers, and technologists who were working to build the structure. As Augustine noted, quoting Frank Lloyd Wright's mentor, Louis Sullivan: "Form

follows function, and function follows form". The architects sought for a synergy between nature's specifications for the biome design and structural criteria. The profound British architectural thinker, John Ruskin, also influenced their work by his writings on beauty and the sublime which he asserts must be united for the greatest architecture to be achieved.

Augustine drew on a combination of sources — poetic impressions, philosophy, engineering and natural history observation as great architects have always done in seeking inspiration for their creations.

"In the rainforest," Margret mused, recalling her field work in the Amazon and Puerto Rico, "you experience an incredible density of vegetation and a cacophony of sounds — mosquitos buzzing, monkeys and primates calling to one another, the songs and calls of birds. Looking up, you are enveloped in different levels of a living canopy — you don't go to the rainforest for star gazing. The rainforest needed to be tall, to allow the trees to soar, and to close in the forest ecosystem. We developed the eighty foot tall stepped pyramid structure to accomplish the broad base moving up to the heights, a fifty foot mountain with waterfall and floodplain for diversity of environments."

Phil Hawes and Margret Augustine standing by a model of Biosphere 2.

93

"In the savannah, you step out of the forest to gaze over great vistas, expanses of grassy fields. We didn't have the miles and miles of Africa or Australia for this terrain, so we selected the main sweep — a long vista for a bonsai savannah, with a stream traversing its length to create opportunities for different habitats."

"The desert is an experience of stark individuals, silhouetted individuals, not great fields of grasses or dense growths of rainforests. The desert has traditionally been a place to go to be challenged, to be tested, to find something within. It is a place of extremes. The forms of the desert — a stark mesa, rocky slopes, low-lying dunes and flats — could be accommodated with a moderate height, amphitheater structure spreading out to the broad flat desert plains. The Biosphere 2 desert is surrounded on three sides by the Sonoran desert outside. The lowest point of all the biomes, the desert will be the biome both the hottest in the summer and the coldest in the winter."

"The marsh is a place of great mystery, tributaries and waterways converging in intricate and mysterious ways, one is drawn further and further in. The estuarine biome needed also a long gradient of salinity, a feature again accomplished in Biosphere 1 over long distances. We developed the twisting and convoluted pathway of the marsh 'stream' bed which provides this gradation without the need for long distances, and placed the estuarine and ocean systems parallel to the savannah separated by the height of a rock cliff rather than miles of space. The ocean steps up from the depths at twenty-five feet, to the teeming life of the coral reef, to the broad and shallow lagoon."

"The intensive agriculture biome (IAB) and human habitat were anthropogenic ecosystems — created by human beings yet calling upon nature. The vaulted structure of the IAB was developed to access maximum lighting for this 'bread basket' of the biosphere where the food for the people would be grown."

"The analog for the human habitat is an urban area, with all the richness in communications, opportunities for technical creativity and science that characterize humankind. The space age design of a curved and rounded habitat was selected to take new advantage of the geodesic form developed by Buckminster Fuller and enhanced by his student and successor, Peter Pearce, our space frame contractor."

Hawes puts it thus: "To move through sculpture is what is happening. We have tried to make the experience of being inside this building close to being a sculptural experience of beauty and function." Besides the vaulting shapes which predominate in the intensive agriculture biome, a major feature of Biosphere 2 is provided by the ziggurat forms at each end of the wilderness biomes. Augustine and Hawes, with their sense of the world's architectural heritage, had been attracted to this stepped pyramid-like form, found in central and south America, early Egypt and Mesopotamia. They decided to have one ziggurat over the rainforest at the northern end and, another to cover the desert down the hill to the south. The two biomes were arranged in this order because of the overall pattern of air-circulation from the rainforest to the desert. The two would be connected by a long section with a pyramidal cross-section, having a rectangular "footprint" — as they called the ground surface of a biome — housing the ocean, the marshes and the savannah.

Micropolis

Initially, many outside observers inquired whether the biospherians would live as a hunter-gathering community, or even more primitively as quasi cave dwellers. The paradigm of an ecological lifestyle had become so dichotomous between the technosphere and biosphere that many assumed that one had to choose between a future that was purely technological or one that rejected technics altogether. The idea of the noosphere, however, called for the harmonious integration of both. The human econiche would be to function as intelligent managers and researchers of the biosphere, not as cave dwellers or science fiction characters in a lifeless world.

The habitat was designed as a microcosm of the life of humanity, a *micropolis*. The micro city of the future would be a node in the network of the planetary communications and information flow, connected to the life of the planet and the human "global village". Biosphere 2 is not an experiment in the isolation of human beings, and careful attention was focussed on maintaining meaningful interaction with the human community and greater biospheric world. Architects Augustine and Hawes reviewed the role of the city in history to evaluate what would be the necessary design functions.

The rise of the city is regarded by many historians as contemporary with, and probably in large part responsible for, human civilization. Throughout history, architects have labored to create new forms appropriate to the evolving life of the city.

In America, Thomas Jefferson championed the concept of the "the country university" — as the center of human life in the New World. In the University of Virginia he attempted to create a new type of architecture—unlike the cluster of buildings of the European university or the cavernous schoolhouses of eighteenth century America. As biographer Dumas Malone described it:

"He aimed for classical, beautiful and dignified instead of medieval,

Architectural rendering of finished biospherian habitat area — a "microcity".

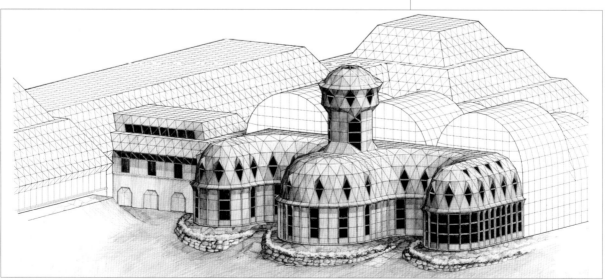

baroque or nondescript; it was to be an aggregation of individual buildings on a hill, spread out in the open country ... In the original plan, the pavilions were to surround the lawn on three sides and to be connected by porticoes onto which the dormitory room opened."

Jefferson's architecture, although derived in part from the Roman villa, sought to express the intellectual freedom of the New World, where individuality and happiness were to be secured by education and free exchange in the 'marketplace of ideas'. Jefferson believed that good buildings helped make good people and that ghettos were responsible for the reverse. A close contact with the natural and agricultural systems which sustained the individual were important, and Jefferson demonstrated this in his Monticello total system plantation estate. The life of the country and the life of the city should be combined to avoid alienation either from nature or from the ongoing development of ideas, art and commerce of the human community. Jefferson also felt that: "Buildings based on mathematics and geometry would always be viable, because they exemplify natural laws". A special trip was taken to Monticello and the University of Virginia by the architects to study what application could be made of Jefferson's ideas and practice.

The Habitat

The spaceframe model used for the design of the habitat portion of Biosphere 2.

T he mathematics and geometry of Pearce's spaceframes revealed wonderful possibilities for the construction of a microcity on organic lines, in keeping with the vision of Jefferson, Frank Lloyd Wright and others. Hawes remarked: "Spaceframe architecture is only about thirty years old and it has mainly been used on forms like domes, cubes or simple pyramids. But, Biosphere 2 is taking two and three dimensional space frames and making them conform to the shapes we want. So from an almost abstract geometrical form, we're taking them and showing the incredible potential they have — like the three-stepped arches of the

intensive agriculture, or the complex ziggurat of the rainforest. Peter Pearce was already engaged in exploring their fantastic potential and together with him we've pushed it even further."

Augustine and Hawes realized the potential of Pearce's spaceframes when they started working on creating a shape for the human habitat of Biosphere 2 — the biospherians' "microcity". They realized they could use Pearce's geometrical capacities to create roomy and wonderful rounded shapes. Spaceframes tend towards spheres and curves because it is the most economical use of materials, and because the lines of force travel better around a curved space than a square corner.

The habitat has its dwelling places, its study areas, its laboratories, its machine shop, its manufacturing area, its medical center, and its lounges for relaxation — more than twenty-five thousand square feet of living space in all. Somewhat unusual about this modern microcity is the inclusion of a barnyard under the same roof, in the style of Old World Europe. The domestic animals occupy the southern lobe of the ground floor in pens abundantly covered with vines, fodder trees, and shrubs.

The habitat is two hundred and forty feet wide and eighty-seven feet high. A circular tower stretches to the highest point housing the library and observatory. Height is one of the dominant characteristics of the arrangement of this microcity — its "verticality," as Hawes calls it. The structure climbs in a series of mezzanines and levels to create an expansive sense of space for the enclosed inhabitants. From mezzanines and balconies on the southern elevation of the building, its residents look out over the fields and plots of their farm from twenty to forty feet below them. Windows on the western and northern exposures look out over the stretch of the Arizona desert.

There were two reasons for this carefully orchestrated height. The lesser was the usefulness of increasing the air volume as much as possible. The more atmosphere enclosed in the Biosphere, the better. The major reason was the aesthetic value of space: to Hawes the beauty of a visual penetration is as important as an actual physical penetration. Hawes speaks with enthusiasm of the eye looking from one space into another, from a low space to a high space, or from "a long, narrow space into a big, fat space." His choice of words reflects the delight he takes in creating an expanded dimension, "creating more with less", as Buckminster Fuller said, by every means at hand.

Hawes, Augustine, and SARBID staff have used every trick in the book to open up the vistas and create sensations of depth in a world that could otherwise become relentlessly claustrophobic for its inhabitants. One example is the one hundred and sixty feet long row of vegetable planter boxes on a catwalk that runs like a veranda along the line of studio apartments some twenty feet about ground level. This row has been dubbed "the salad bar". A person in one of these apartments can look out through the glass wall, see a row of tomato plants and carrot tops on the catwalk, close-up, and then gaze out across the farm below. The wall on the far southern side of the Biosphere seems appropriately distant. It's the kind of diversity that Hawes and his staff have struggled to achieve for every perspective: juxtaposing the horizontals against verticals, close planes against a sequence of more distant planes, the play of shadows, the texture of surfaces.

A Multidimensional World

The biospherians will live in a multidimensional world of different shapes, varying levels, and winding walkways. Staircases dropping down or climbing up to the next level or mezzanine break up the monotony of a homogeneous space. One delightful surprise is a bridge leading from the analytical lab across the main corridor to the medical lab. Walls and ceilings often curve, as do most of the staircases. One of the handrails on the gently curving staircase by the analytical lab is molded of clear plastic.

A rosy flagstone paves the walks inside. A multitude of potted plants populates every conceivable nook and cranny. The potted plant has taken on a new significance here. Aesthetic decorations, they are now also gainfully employed as air purifiers and needed biomass.

Walls are white but color abounds, not only in the plethora of plants, but in the furniture, wall hangings, and carpeting. The Wool Bureau, a sponsor for the natural fiber products used in the habitat, advised on the interior where fabric and carpeting were needed. This provides acoustic dampening as well as textural variety; no one wants a noisy echo chamber outside their apartment door.

The biospherians' apartments measure approximately twelve by sixteen feet with a mezzanine area bedroom overlooking a sitting room below. Each room has a view; those that don't look out over the farm, look out across the Arizona desert.

Almost everything is open to light: light pours into the main corridor from its row of tall narrow windows, into the kitchen area from the wall of glass above the orchard, into the analytical lab from skylights, and into the eight foot stairwell in the tower that ascends to the library through its twelve foot high diamond-shaped windows. Hawes has described one aspect of this abundant light: shadows of space frames will break up the surfaces of the even walls, blue and white patterned sky overhead reflecting and casting sunlight into corners, creating a moving collage.

On the northern side of the habitat is the "command room," which houses the biospherians' equipment for contact with the outside world. Hawes wanted to avoid the stark monotony of the command room at the Johnson Space Center where rows and rows of computers and monitor screens restrict movement and comfort. Here the round command room emphasizes the verticality, with a high ceiling allowing a row of tall, narrow windows that look out onto the slope of desert rising up to the north. There's generous space for the biospherians' own individual work stations, as well as for all the instruments and computers.

For exercise the biospherians have an area on the second floor facing the IAB. Some of the staff expect that the crew won't need to use the exercise facilities very often, as they'll get plenty of exercise elsewhere. A biospherian could easily walk five miles a day just going about his or her daily farming, maintenance, and field observation duties.

Although designed as a prototype for a long-term Mars colony, the

Model of the interior stairway to the tea room and observatory on top of the tower.

ROBERT HAHN

feeling is not at all the cramped, clinical enclosure portrayed in science fiction. It has the feeling of a comfortable, livable community, achieving what astronaut Rusty Schweickart, at *The Human Quest in Space Symposium* in 1986, probably had in mind when he declared that the true challenge in space "is not simply existing or surviving in space, but *living* in space; to have the opportunity for someone in the future to take a picture of two kids walking hand in hand or of a mother and child, not living inside a spacecraft, but living in a community in space."

8
The Assembly

"We all know the saying that a chain is only as strong as its weakest link."

Buckminster Fuller

Management

Most large construction projects have a general contractor, the big boss, who in turn hires subcontractors to carry out different parts of the project. Subcontractors usually focus on their particular jobs, looking from the bottom up, with the boss at the top. The structure is hierarchical. The SBV staff felt such a structure in this case would make conflict too likely.

At another construction project in New Mexico, Allen, Augustine, and Hawes had experimented with their construction crew; short hours and a short work week, but hard work, good wages, no drugs or alcohol, frequent job-sharing, and non-sexism. For example, they insisted on equal numbers of men and women on the construction crew. They also insisted on self-reliance and self-motivation. That idea could still generally be considered "advanced" today, but in 1975 it was considered revolutionary.

Augustine wanted to build Biosphere 2 using a similar form of management. That meant no general contractor. Instead, there were a dozen or more contractors who each had complete responsibility for a part of the project. They were expected to work together and integrate their tasks. Candidates for the eight posts as 'biospherians' held key responsibilities of their own, and worked closely with the contractors. Bernd Zabel, a mountain-climbing electrical engineer from Munich, West Germany, was put in charge of managing construction of the structure. Mark Van Thillo, a Belgian machinist and tool and die maker who had spent two years as ship's engineer aboard the Institute of Ecotechnics' sailing ship, the *R/V Heraclitus*, was in charge of the day-to-day flow of the work and quality-control.

C. ALLAN MORGAN

Biosphere 2 space frame construction, November 1989.

Opposite: Spaceframe construction workers, secured by safety belts, crawl over the structure like spiders.

C. ALLAN MORGAN

The team of Biosphere 2 construction managers in March, 1990.

101

PHIL HAWES

Biosphere 2 construction site at sunset.

Both would remain inside Biosphere 2 when the door was shut and sealed. Zabel was fond of telling people that he was truly dedicated to his work in supervising the building of Biosphere 2 because he would have to live with the consequences more than anyone else who had ever supervised a large construction project. If the roof leaked, pipes broke, or electrical outlets shorted out, there'd be no getting away from his most immediate critics, the other seven biospherians.

In 1986, both these technical experts were given the kind of challenge many people would love to have the chance to face: the building of Biosphere 2. Neither Zabel or Van Thillo had any illusions about the many caveats they would have to deal with or the intense daily grind that lay ahead to get the mission accomplished. They had an army of people to bring together, people with widely divergent tasks and backgrounds.

From the start, the two men developed a management strategy that allowed the contractors to manage their crews and work out their own problems as they arose, so long as the critical path was not affected. They would not play the role of mediator, continually resolving arguments and stopping fights, unless there was a spill over into another contractor's area.

Such a system for people accustomed to either being boss or being bossed was not always easy to maintain. As construction proceeded, conflicts began to crop up more and more frequently, particularly as things grew more complex on the site and some of the contractors began to run into each other trying to do a different job in the same place. Zabel and Van Thillo would "call down the critical path chart" to establish the priority and ask the subcontractors to work out their schedules accordingly. They worked from the top down — maintaining the vision of completing the whole helped work out the individual parts.

The encountering of differing corporate cultures produced excitement at times. Zabel and Van Thillo found they had to go on a fast learning curve to deal with the complexity of thirty-four subcontractors, each of whom had different quality standards to meet, nearly all of them higher than in previous jobs. "Why is all this extra care really necessary," some asked at first.

But significant signs soon appeared. In the early days, workers taking last drags from their cigarettes would flick them down into the soil. Van Thillo, walking through the construction zone on his inspection tours, would spot the butts, pick them up, and either pocket them or toss them over the concrete walls and out of Biosphere 2 into trash bins. The workers got the point and the number of butts disappeared, and morale went even higher.

As the substructure of Biosphere 2 was completed, trucks began hauling in loads of fill and dumping them around the site. They all looked like plain old dirt to some people but were actually carefully selected components for making soil. "No, not dirt," Van Thillo corrected a visitor. "It's *soil*." These piles would comprise the substrate layers for each of the biomes, each of which would have different soil compositions. In general, "basement soil," for the bottom layer was gravelly and brownish. On top of it went purplish soil, rich in potassium, calcium, and sodium. Next came soil containing considerable limestone mixed with compost, high in calcium carbonates. It was a grayish pink. Twigs were mixed with the soil to improve drainage and add organic material.

JOHN CANALOSI

Biosphere site desert soil move-in, January 1990.

"That's not dirt down there," a glazier tethered to the spaceframe above the agricultural biome reminded one of this fellow workers one day. "That's a farm."

The relationship between SBV and the contractors began to improve when Zabel and Van Thillo took them out for a night on the town. They started with beers at the Lariat Lounge in nearby Catalina and then, heading on into Tucson, stopped at the Bull Ring for dinner. All the while, Van Thillo and Zabel were trying to impress the construction supervisors with the magnitude and importance of the project. Before the night was over, Al Schroeder of Diversified and the other construction guys, trying to see if Van Thillo and Zabel were for real, led them further into town to one of the topless bars along Miracle Mile, where an MC, learning of their presence, flipped on the PA system. "I hear we've got some biospherians in our crowd here," he announced. "Let's welcome them with a round of applause." The contractors had all done excellent work up until then, but from that point on tempo and morale were also tops.

Construction

Astheoutsideslowlytookform,sodidthemechanicalandartificial systems inside. Work had begun on the wave machine which would be needed for the miniature ocean. In the desert biome, the basalt "rock" formations were completed. Painted-on lichens covered boulders. "Rocks" for the wall along the ocean were sprayed on as wet cement, troweled to form shapes and, when dried, painted. Workers added pockets for soil. The cliff and other outcroppings were almost complete. Styrofoam blocks jutted out of different parts of the rocky walls. After the artificial rock hardened, the blocks were removed, leaving planting holes from which vines and branches would eventually trail out and down over the mountain structure.

Sarbid Architecture designed and Larson Construction Company built the great cliff face plunging from the savannah to the ocean; it provides a number of econiches, particularly for the lizards.

Other natural features simply couldn't be hauled in or created overnight. The rainforest mountain was especially complex. Its infrastructure is much like a parking garage: square on top, constructed of steel-reinforced concrete. It's hollow interior serves as one of the major ducts for circulating air throughout the rainforest. It also has space inside for storage of tools, and for the machine that will force water under high pressure through fine nozzles to create the cloudforest fog. Outside, once the surface was covered with concrete sealant, the mountain facade was ready to receive a rock face, shaped to imitate the forms of nature, but built from special concrete.

The shape of the cloud forest plateau was designed based on a Venezuelan sandstone mountain.

The Larson Construction Company in Tucson, specialists in creating natural forms out of concrete, had built rock-like structures and other concrete creations at the Arizona-Sonora Desert Museum, the Tokyo Bay Aquarium, and Universal Studios. At Biosphere 2 they began with a model of the mountain and photographs of the Venezuelan sandstone outcropping on which the mountain was based. Photographs also guided them in the creation of granite boulders, lava and basalt in the desert, and limestone cliffs near the ocean. Larson first built a skeletal form using three-eights inch steel reinforcing bar, a weight that allows easier bending when creating the curves, crags, nooks and crannies, overhangs, ledges, and openings of mountain surfaces. A layer of concrete — a rich recipe rated at four thousand pounds per square inch, a third stronger than what is typically used for house foundations — was sprayed pneumatically onto the steel rebar skeleton, forming a rough, mountain surface, four to five inches thick. After curing, the Larson crew laid another inch of concrete and sand, then carefully troweled it to the texture and shapes of the natural rock. Taking up their brushes, they then painted the surfaces with acrylic vinyl water-based latex paint, reproducing the charcoal face of a Venezuelan sandstone mountain.

To the east of the mountain, down in the lower areas of the rainforest, one of several air shafts poked its square nose up into the air. The rock would disguise these shafts as small caves. Giant fans in the basement would suck air down into some of them, and blow it back out through others.

Southeast of the lower rainforest, down beyond a cliff of artificial stone lay the site of the ocean. Tractors and trucks brought loads of carefully composed rock and sand to a continuously moving belt which dumped it where the ocean would be. A giant crane had lowered a small dozer down onto the ocean floor, where it crawled about, spreading limestone substrate for the ocean bottom. A concrete ridge rose up from the bottom, and huge limestone rocks were clustered along it. This would be the location for the coral reef. The reef would sprawl along several

JOHN CANCALOSI

underwater ridges and humps. Dempster had calculated that the force of the wave machine might sweep away ridges made from rocks and sand, which is why it was necessary to make some of the reef emplacements out of concrete, and top them with rock.

Soon after the beach was put down, the filling of the ocean began. The ocean recipe called for approximately ten percent seawater trucked in from the California coast near La Jolla, and the rest freshwater, flavored with a commercial recipe of salts and seawater ingredients called "Instant Ocean".

Introduction of soils in January 1990. There are forty-five different soils in Biosphere 2. Others will evolve.

Abigail Alling, biospherian and Director of Marine Systems, took control of preparing the site for the arrival of the ocean water, the coral reef, the fish, and other components of what SBV, conscious of the Sunspace Ranch's four thousand foot altitude, was fond of calling "the highest ocean in the world".

By early spring of 1990, she was facing a bottleneck. The reef had to be raised from the Caribbean and brought to Arizona before late July, when the hurricane season would force the *Heraclitus* and another support vessel, the Smithsonian's *Marsys Resolute*, to seek safe harbor. She needed about six months to get the ocean water in and stabilized for reception of the reef; reefs are sensitive to the nutrient levels of water and the amount of oxygen in it. She had only three months. Soon after the water was put in, there was a bloom of algae. Alling and Adey had expected this might happen, but even so, it would be a battle to overcome it. The scrubbers would slowly clean the ocean of the excess algae, but it would take time. With time running short, she and Adey had to shorten the process somehow without jeopardizing the survival of the coral.

The wave machines and the algae scrubbers also were problem-plagued. A large partition separating the ocean from the marsh was a component of the wave machine. It was hollow, forming a large holding chamber for water. Vacuum pumps would suck out the air inside the chamber, pulling water into it through a long, underwater opening. Every eleven seconds, the vacuum would release itself, sending a gush of water back through the opening, producing a wave that would roll across the ocean.

Some of the engineers feared that the constant flow of water into and out of the chamber would eventually erode the concrete, just as the flow of rivers can cut through rock. If this erosion were severe enough, the wave machine would shut down, unable to maintain a vacuum or hold a sufficient volume of water. That would sentence the reef to death, as it needs the constant wave action to circulate nutrients and oxygen. They finally decided they could protect the concrete with a special paint, but it involved a laborious process causing more delays.

GILL C. KENNY

Left: Divers plant Biosphere lagoon, 5 October 1990.

Below: Abigail Alling with fan coral collected in the Yucatan destined for Biosphere 2.

MARGRET AUGUSTINE

By April, Alling had made progress. The ocean was full of water and she was regularly donning scuba gear to take a look at its progress below the surface. By late May, she had turned on the wave machine, which was pushing foamy breakers onto the white sand beach.

Pacific Aquascape, the contracting firm, had designed vacuum pumps that were so gentle on the seawater that a fish could conceivably ride from the ocean through one of the algae scrubbers and back again without harm. The scrubbers were now online, cleaning the seawater. That May two huge trucks arrived, bringing in the first of the algal turf and grazing fish prior to the arrival of corals themselves. The marsh was also installed and thriving by mid June, with bugs whirring and frogs hopping among the mangrove roots and salt grasses.

On May 24, 1990, the last section of spaceframe had been lifted above the rainforest and set into place, topped by a small swaying pine tree and the American flag, a Pearce Structures tradition. In the Arizona desert, under the shadow of mountains, a new world was taking shape.

Collections

In 1988, various teams converged on the Everglades outside of Goodland, Florida, to begin the process of collecting marsh plants for Biosphere 2. Roy Malone Hodges, manager of the marsh biome and in charge of logistics for its move-in was there, as were Van Thillo, Zabel, Alling, and Rodney Solomon, SBV Logistics Director at that time. Adey and his capable crew from the Smithsonian Institution also came down to the rendezvous, six to eight person teams in twenty-one foot Boston whalers. The Institute of Ecotechnics' ship, *R/V Heraclitus*, also sailed in, under the command of Expedition Chief Goga Malich, to assist the collecting.

SBV didn't immediately transport the marsh modules the teams gathered to Arizona, but kept them in Florida until Biosphere 2 was ready to receive them. The collections were maintained by connecting the over one hundred four by four foot wooden containers with six inch pipes through which water was pumped to simulate the tides.

Beginning in December 1989, trucks loaded with the wooden containers of mud, big scoops of marsh plants, and microbes flooded with water from the marsh, began to leave Florida. The caravan crossed Georgia, Alabama, Mississippi, then Louisiana and Texas — all states with marsh habitats similar if not identical to the Everglades. New Mexico, however, was foreign territory for marshes, and the trucks ground to a halt near the Arizona-New Mexico line, near the Arizona town of San Simon. An agricultural check-point screens vehicles crossing the state line to prevent destructive insects, noxious plants, and other invaders that pose threats to agricultural crops invading Arizona. Confusion began when an official mistook the word "mangrove" for "mango"; understandable as "mangroves" wasn't even on the list of things one might conceivably be trying to bring into dry Arizona.

JEFF TOPPING

Above: Roy Malone Hodges measures salinity and temperature of the marsh.

Right: The marsh had to be assembled in a large temporary greenhouse in preparation for the round-the-clock introduction into Biosphere 2.

C. ALLAN MORGAN

C. ALLAN MORGAN

Hodges soon found out that officialdom is the most difficult part of logistics. Alling and Leigh worked continuously with inspectors on the permits, while Hodges went back and forth to the border. Nonetheless, the Arizona officials struggled admirably with plants and animals no field guides had ever been written for in marshless Arizona.

The mangroves propagate by shedding seedlings that look like round lobes. These float in the tide until they lodge in soil, take root, and grow into new trees. The Arizona agricultural inspectors plucked such a nodule from one of the trucks and sent it to a state laboratory in Phoenix for inspection. For the next four days, the mangroves sat on the Arizona line, waiting.

Word finally came back that all was clear. But that was just the beginning. In the end, twenty-five truckloads of marsh collections had to make their way across the country and through San Simon border station to pass inspection.

The idea of hauling large chunks of marshland, of whose contents only the major species were known, or even visible, perplexed inspectors who were accustomed to seeing truckloads of single species. SBV wanted the plants and all the bugs riding on them. The inspectors were used to making sure "bugs" didn't get through. Hodges found himself spending hours on end picking spiders off of plants and uprooting grasses classified by agricultural experts as "undesirable".

At times, Hodges said, inspectors required him and his workers to spray for hostile insects, a tactic SBV tried hard to avoid, both to preserve the natural insect fauna and to prevent the potential build up of artificial chemicals in Biosphere 2 that could play havoc with the food web and the health of humans, animals, plants, and microbes alike. Ultimately, when it was clear there was no alternative, they used insecticides, but chose varieties that broke down to natural components when

Top: The algae scrubber system backs up the ocean's own purification systems.

C. ALLAN MORGAN

Above: To maintain the marsh mangrove system while waiting to be moved into Biosphere 2, the boxes had to be all piped together into a small body of salt water which simulated the tide.

exposed to ultraviolet light or which could be readily removed by rinsing.

Roy Hodges, his high morale truckers led by John Grady, and others at SBV spent countless hours courting, persuading, explaining, fraternizing, and jawing with agricultural inspectors on the differences between saltwater marsh and Sonoran desert ecology. They invited inspectors to Oracle to tour the Biosphere 2 site, to give all of them a clear idea of what the project was about. The inspectors also helped SBV set up its quarantine area, a requirement for bringing many of the plants and animals into Arizona. At first concerned about "swamp things" getting into Arizona, the agricultural inspectors warmed to the project and SBV soon enjoyed a strong and cooperative working relationship with them.

ROBERT HAHN

Plant donations also came in from the Missouri Botanical Gardens, Fairchild Botanical Gardens, and New York Botanical Gardens, some of which struck deals with SBV to exchange seeds from their gardens for some of those SBV was collecting from around the world. Greenhouses with temperature and humidity controls were constructed to propagate the seeds or to hold plants that had already been brought to Arizona. The task required a newfangled contraption or two, such as the "tide maker", a mass of pipes and pumps that kept water rising and falling around the Everglades mangroves, a necessity for keeping them alive until they could move inside Biosphere 2.

These specimens were contributed to the Biosphere 2 rainforest by Missouri Botanical Gardens thanks to its Director, Peter Raven.

The tropical plants, accustomed to high levels of humidity, found their Arizona homes equipped with artificial fog makers, which periodically produced a consistent mist in the air around them. This dependence of plant on machine brought into full bloom the differences between engineers and botanists. When the fog machine went haywire one day, a band of engineers and technicians marched down the hill to the greenhouse and discovered that several varieties of tropical vines had grown into and intertwined themselves with its mechanisms. For them, the solution was clear. They hacked away at the plants until the machine was free. The botanists were outraged. By talking through such experiences, the engineers and the biologists came to understand each other better.

Warshall planned to model a portion of the Biosphere 2 savannah after the Rupununi savannah in Guyana, and in spring 1988 he and Leigh, accompanied by a film cameraman, left for South America to get their grasses. After surmounting the expected unexpected difficulties, they eventually reached grasslands and, within about a week, the SBV greenhouses in Arizona were teeming with dozens of species of South American savannah grasses, all duly quarantined by federal and state agricultural officials.

Ecotechnic Developments

State and Federal regulations in some instances called for mandatory application of certain, usually pretty nasty, chemicals to imported plants. What could be done with these substances afterward? Some of the engineers suggested that SBV handle it like everyone else did — take the poison down to the University of Arizona toxic waste dump. But that would hardly be in keeping with the spirit of the enterprise, nor be a decision one could in good conscience live with. The solution was to

The wilderness biomes in Biosphere 2 began to take shape by late summer of 1990.

This photo is of nearly the entire construction group on work at the site in the summer of 1990. Their constantly high morale and quality work was a a major boost to Biosphere 2.

make this problem, one not unique to this project, an opportunity for a useful and commercially valuable technology development. The Environmental Cleanup System (ECS) was researched, designed, and put into operation. ECS moves wastewater from the quarantine area by a conduit through a sophisticated system which permits great flexibility in the way chemicals can be neutralized. Then they are subject to bioassay by exposure to sensitive marine organisms, which will verify that the effluent is safe to discharge into the environment. This system could have wide-ranging applications in monitoring the quality of water, detecting pollution and cleaning lakes, rivers and perhaps even oceans of untreated toxic pollutants that have been discharged into them into the past.

C. ALLAN MORGAN

This is one example of how ecotechnical developments for Biosphere 2 will have wider application. Not only will ecotechnics help us to tackle environmental problems, but it can also help to build new environments, exotic worlds in unusual places for research, education, and for public enjoyment. Perhaps one of the most popular products to come out of the Biosphere 2 project will be such "ecoscapes" — landscapes with emphasis on a given ecological region — for use in interior landscaping or in parks. Robert Hahn, SBV's Marketing Director, has high hopes for this kind of business: "We've amassed tremendous knowledge in soil, rocks, geology, morphology, what plants to use and why, as an outgrowth of our work in designing Biosphere 2's wilderness biomes. We now know how to build rainforests, savannahs, indoor oceans, and deserts."

9
Cybernetics

Breathing and heartbeat diminished, concentration intensified. It seemed to me that something extraordinary in the forest was very close to where I stood, moving to the surface and discovery. I focused on a few centimeters of ground and vegetation. I willed animals to materialize and they came erratically into view ... together they composed only an infinitesimal fraction of the life actually present. The woods were a biological maelstrom of which only the surface could be scanned by the naked eye.

E. O. Wilson

Dialogue

Perhaps the most interesting aspects of Biosphere 2 is the way that it "talks to" and exchanges information with Biosphere 1. Humans in two quite different metabolic systems will now be exchanging words, numbers, sounds, images, procedures, patterns, at rates too quick and on a scale too vast to comprehend. No one can predict what the outcomes of this massive exchange will be. One effect already is the sense of an emerging *noosphere*, a new world of intelligence—intelligence used in the original sense of the word, from the Latin, meaning the ability to learn, or understand.

The work that went into the cybernetic system of Biosphere 2 pushed the limits of computer science into the realm of artificial intelligence. This new state-of-the-art allows for a flexibility within the monitoring system of the biosphere and technosphere which can evolve with Biosphere 2's management requirements. But how will humans interface with the world of computers and computers with the world of life? Who will be in control?

Control

SBV was on the verge of launching its own spaceship, so to speak, with eight people in it. It needed at least a rough draft of an operating manual. Like watching the gauges on a dashboard, those inside and researchers on the outside needed a way to monitor the vital signs of the system and take action — from routine maintenance to emergency measures. This would be the kind of information that the biospherians would use to decide, for example, whether they should cut back one species to prevent the death of another more crucial species. Or when they might need to intervene to avert CO_2 overload.

The task of monitoring Biosphere 2 information might be so overwhelming that the ability to process and use it would be hampered. The sheer quantity of information make it difficult for the biospherians and Mission Control to understand what was happening and take meaningful action. If you have thirty-eight hundred species to deal with, how can you keep track of the effect of one of them on another in light of the fact that thousands of others are also having their own effects at the same time?

One idea was to build two Biosphere 2's side by side, one as a control

JEFF TOPPING

A view of the Biosphere 2 "Nerve Center" of Mission Control where the Biosphere 2 "nerve system" was developed and which will be in charge of maintaining communications with Biosphere 1.

Opposite: Noberto Alvarez (foreground) director of the development and application of the SBV five level computer system that provides tremendous intellectual power for the eight biospherians to use in managing their enormously complex world. Tom Kettler (left) and Francisco Luttman (right) are SBV members of the team. Consulting contractors include Hewlett-Packard, Systems Integrated, Gensym and Oracle.

for the other. But this meant virtually doubling the expense, and was out of the question. If an ecological balance occurred inside and eight biospherians and the life around them sustained themselves, then everyone could get down to studying the precise details of *how* it worked. After all, nobody was calling for scientists around the world to stop studying Earth's global ecosystem just because they didn't have another planet identical to it for comparison. In another sense, the construction of Biosphere 2 was the construction of a control. Biosphere 2 presented the opportunity to develop a comprehensive biospheric monitoring and management system on a scale more approachable than Earth's. The successful development of such a model would be a large stride toward developing a more accurate model for Earth.

"If it doesn't work, if we can't balance the atmosphere, we can go back in and try again," said Tony Burgess. "I guess the Wright Brothers had no real controls either."

The Naturalists

Although it is artificially assembled, Biosphere 2 depends on the most basic of natural phenomena. And its human members modelled themselves on a variety of roles, from the gardener, to the naturalist, to the wildlife manager. Like those role models, the biospherians' perceptions would depend on their senses of smell, sight, taste, hearing,

Linda Leigh examining the state of the grasses in the lower savannah.

and touch. For example, a trained gardener will notice fine gradations in the nutrient or water needs of his plants as he walks past them; and he can tell much by the look or smell of the soil. Likewise, a naturalist in a wilderness can discern subtle changes in the environment, or track the activity of particular animals and insects.

Edward O. Wilson, the Harvard entomologist, termed this type of human attention 'the naturalist's trance.' The naturalist would react with a basic instinct, the kind of sense registration an experienced observer gets walking through an agricultural area; perhaps trimming leaves that require pruning, overriding irrigation patterns if plants and soil look and feel too wet or dry, noting conditions of plants by their color, making mental notes of flowering patterns, the development of fruit, anticipating schedules for ripening, and making plans for harvesting. In a wilderness area, the naturalist would document animal populations by counting them or making estimates based on nests, droppings, or frequency of their calls. The naturalist would anticipate that a drought affecting growth of eucalyptus, for example, would forbode possible starvation and decrease among populations of koala bears.

On a higher level, the naturalist looks at the *patterns* of the biome, rather than giving exclusive attention to individual plants and animals. For example, the astute naturalist would note short-term and seasonal patterns of insect movement and where they cluster. After long-term observation, comparisons between years would become obvious. If the desert began going dormant a bit earlier, was this a function of a change in watering regimes, night air temperatures, or their combination? Or something else entirely? Whiteflies generally attack squash first and only then move to cabbages — but what if they had changed their modus operandi? To be an effective naturalist at this level requires training and experience. For a veteran rainforest ecologist like Prance, the complexity of discernible patterns is great because he has seen rainforests at all seasons and in all conditions — pristine, secondary, disturbed, regrowing, or replanted. Such a relationship between its humans and their surroundings could be essential in helping Biosphere 2 reach a sustainable balance.

Naturalist observation has served humans well, from the beginnings of agriculture to the beginnings of the modern environmental movement. The absence of the trills, warblings, and staccato of songbirds led Rachel Carson to notice the effects of pesticide spraying she documented in *Silent Spring*. Sometimes, the health of a single species can speak for an entire ecosystem, such as the giant clams of the ocean, which indicate their state of health through the degree of beauty and richness in their colors.

Besides watching their own health in Biosphere 2, humans would keep an eye on indicator plants — those first to show nutrient or pH stress — and also on 'keystone' predators — including the humans themselves — whose continued presence and health indicates that those links supporting it in the food chain are in good vitality. For example, keystone predators, which are at the top of the food chain, may stop reproducing because of a break in the food chain below. But some physical characteristics are simply beyond the range of human senses. Trace gases, for instance, would have to be monitored electronically or by taking samples and processing them in Biosphere 2's analytical lab.

Sensors

Despite their limitations, computerization and sensor nets could take some burdens off of the biospherians and act as an early detection system. The needs of what came to be called the *nerve system* were defined by a group of ecologists and engineers to identify the types of information which would be of most value in determining the health of the ecosystem. Collected with sensors, these and other physical data will be fed into computerized databanks, where programs or programmers can retrieve them. Discussions led to the development of more than two thousand electronic sensors for placement around the interior of Biosphere 2, some of which give several different types of information.

Early in the project, a few computer consultants advanced the rather novel idea of branding bar codes on the animals so they could be traced like checkout items at the grocery store. That idea actually had been used elsewhere to study the comings and goings of individual bees from a hive. But movement of animals wasn't an indispensable indicator of the health of an ecosystem, at least not important enough to make this practical.

Sometimes, new sensors had to be developed. For example, how can you measure the temperature just below the surface of a leaf? This required a sensor about the thickness of a fine copper wire, suspended just below the membrane of a leaf surface.

Continuous temperature sensors take readings every ten seconds, inscribing them on a strip chart recorder that can play back episodes of

Some of the computer data banks.

Biosphere 2 like a movie. For routine monitoring, however, Mission Control — the computer center — can use average readings, perhaps four an hour, calling up the ten second readings only if an unexpected fluctuation calls for a more detailed look. The sensor system includes 'sniffers', tubing which will lead air from six locations to the analytic lab where concentrations of carbon dioxide, oxygen and trace gases (carbon monoxide, nitrous oxides, sulfur dioxide, methane and other hydrocarbons, hydrogen sulfide, nitrogen oxides, and ozone) can be monitored.

The leading challenge in developing the analytical lab was finding ways to avoid the need for chemical solvents that would pose disposal problems inside Biosphere 2 as they are highly polluting. The ingredients for standard analytical chemistry include methylene chloride, acetonitrile, hexane and trichloroethane, and other toxic substances which are ordinarily needed in large quantities.

Early discussions about the analytical lab had been divisive. Only after much debate did Hewlett-Packard, the main consultant and contractor, agree to actually put the lab *inside* Biosphere 2. There was not only the problem of releasing chemical reagents into the enclosure: repairs would be far more difficult to make and it would be impossible to upgrade any of the equipment during a particular closure experiment. Taber MacCallum, biospherian and head of the analytical lab, was adamant: the lab had to go inside because the whole project depended on extremely fine and real-time analyses of air and water conditions. His determination at this point led him to find and develop the most advanced systems anywhere. He had in the forefront of his mind the fact that this same problem would have to be dealt with for a Mars or other space project — the sooner the challenge was met, the better!

Taber MacCallum, Director of Analytical Systems and biospherian, contemplating what he calls "the so many variables".

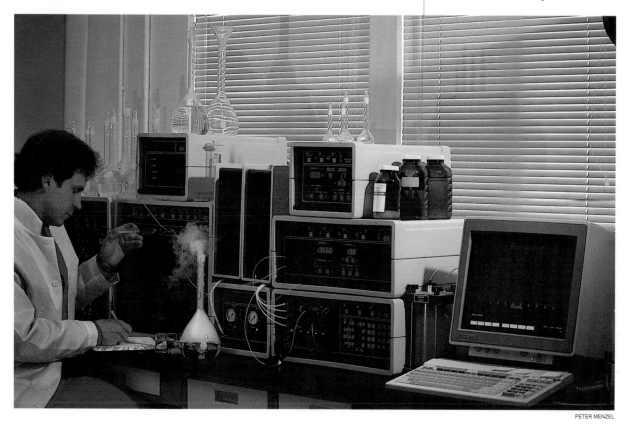

PETER MENZEL

The Nerve System

All of the data from the more than two thousand sensors, as well as the data generated by the analytical lab, will be stored in a computer archive of Biosphere 2, accessible from both Mission Control and Biosphere 2. Like information from Moon shots, interplanetary probes, and ground-based telescopes, scientists will be able to retrieve the information years after the events for analysis. Intelligence, human assisted by artificial, in turn will interpret it and decide on responses, and then watch for feedback to see if the responses work. The biospherians will access the information from the system via the computer screens — Biosphere 2 will be a virtually paperless society.

In a living organism the nervous system does the job of collecting information and making sense out of it. Biosphere 2 needed a "nerve center"; but, when SBV began putting together the computerized elements of Biosphere 2 in 1985, programmers under the management of Norberto Alvarez-Romo, SBV's Director of Cybernetic Systems, decided that a distributed network of personal computers, or one with no central brain, would work better because the design was less prone to the paralyzing "crashes" that sometimes plague central computers. It was cheaper, as well.

The nerve center will sound alarms automatically if carbon monoxide, for example, rises to threatening levels. It will turn on fans and air coolers or heaters to prevent the rainforest from ever going above ninety-five degrees or below fifty-five. Built into its artificial intelligence are the rain requirements of each biome by season, irrigation needs of the farm area, the expected condensation of water as air passes through the air handlers, and other details of Biosphere 2's water cycles. Pumps will turn on and off to help move water back uphill in the course of its movement through the system. Probes in the soil will keep track of the moisture content of the soil and compare it to the desired levels.

One of the over two thousand sensors in Biosphere 2 that will allow for the first time exact measurements of the key variables of an entire world of life. These numbers can then be modelled, among many other applications, to make predictions of Biosphere 1 behaviors.

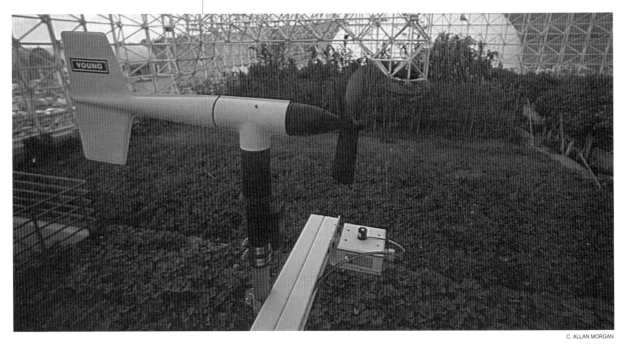

C. ALLAN MORGAN

Water Control

At any moment, controllers in Mission Control or in Biosphere 2 may call up a screen to check the state of the water systems: irrigation, stream, marsh, or seawater. In the marshes, sensors for salinity will maintain the gradient, from fresh-water marsh to ocean. From the ocean and marsh, water will flow through pipes to the scrubbers. From these pipes, samples will automatically be taken and analyzed for nutrients.

The degree of complexity is formidable. For example, the water system requires measurement and action to manage the following elements:

- water to fulfill the rain requirements of each biome; this includes variations for individual ecosystems within the biome, as well as seasonal variations
- flooding requirements for rivers
- irrigation needs of the intensive agriculture system which varies according to crop rotations and season
- condensation loss of water from atmosphere in air handlers and other technical systems
- natural condensation
- actual soil moisture content vs. desired levels
- humidity levels
- levels of storage water in various holders—potable, irrigation, seawater and sumps
- flow of water in pipes and contents of pipes
- tidal flows in ocean and marsh
- water flow to nutrient scrubbers from aquatic systems
- water quality assessments

The water system also maintains a "balance sheet" taking into account the global water use in Biosphere 2 over periods of several days. This monitoring capability allows the system to make adjustments for situations, for instance, in which there is insufficient water to supply the rainforest on a given day, but sufficient over a two-day period. The nerve system can carry such water debts forward, and determine how long such a need shall remain valid before a reassessment is required. The system also takes into account the quality of the water available and being asked for.

Emergency Response

All of the mechanical systems of Biosphere 2 and all of the decisions on selection of plants were intended to result in a single system that would sustain itself — including eight humans. It was understood that in developing such a complex system there would be errors. SBV turned its attention to identifying potential failures and finding ways to fix them.

In John Allen's mind, a skyrocketing carbon dioxide level, catastrophic species extinction, serious problems with the sealing, and a shorter life span than expected were the things he feared more than others. Allen

Reflections of the spaceframe in a pool of water along the savannah stream.

121

GILL C. KENNY

A specially trained Biosphere Emergency Response Team goes into action when needed.

thought the longevity of Biosphere 2 was really the greatest challenge in the long-run, but problems there may not show themselves for years. The most important thing in the short-term was correctly to make Biosphere 2 sustain itself.

Because stability would be for nought if Biosphere 2 wasn't sealed, quick response to a break in the seal became extremely important. Though it's hard to conceive of a broken window in Biosphere 2. Dempster is confident that the glass is virtually impervious to anything up to rocks the size of cantaloupes. He has an unswerving faith in the resiliency of the laminate. Van Thillo shares his opinion. Early in 1990 he witnessed a glazing team accidentally drop a pane from a height of thirty feet. It slammed into the ground, bounced, hit the ground again, and rattled about like a hubcap before coming to rest with nary a hairline fracture.

But to have a plan in place in case of the improbable break or some other unforeseen hazard, SBV organized a team that is ready to take quick action to reduce the loss of atmosphere and then repair the damage. A substantial leak would quickly deflate the lungs, because of the weights on top of the two diaphragms. To reduce loss of atmosphere, in such a situation Mission Control would immediately turn on fans strategically placed in the lungs to neutralize the mild positive pressure produced by the weights, equalizing the pressure inside and outside and minimizing the exchange of gases between the two.

Crew members would climb onto the space frame, tether themselves at the site, and cover the hole, with a sheet of plastic and tape designed to allow as little air as possible to pass through the adhesive. Zabel's mountain-climbing experience would come in handy. Then a new pane would be installed from the outside. "A crew on the outside would have to be ready at a moment's notice," said Van Thillo, who, as the mechanical systems troubleshooter, would lead this kind of repair action. This kind of emergency squad has been informally dubbed BERT, for Biosphere Emergency Response Team.

Biosphere 2 has redundancies built into it that are intended to minimize the effects of mechanical failures. The ocean system, for example, was designed so that any water escaping from the pipes that fed the algae scrubbers would flow to a sump pump which would automatically switch on and return the water to the salt marsh. If the water had picked up debris or contaminants, the marsh's natural tolerance for impurities and its purification ability would take care of them. The marsh would protect the ocean from poisoning. Getting the water back into the system quickly was important: a drop of about a foot in the level of the ocean would prevent the wave machine from making waves, and the coral reef ecosystem — dependent on the pulsating water — could collapse in less than twenty-four hours.

Consistent temperature control is another high priority. The first and simplest level of control is to keep the air moving. In the wilderness, for example, seventeen airhandlers perform the task. Even on the hottest days, all of them wouldn't be needed. The prospect of one or two of them being down for a few days is no cause for concern. The probability of a breakdown of ten or more, based on the record of reliability of the airhandlers, is exceedingly remote.

Energy

A view of the Energy Center through Biosphere 2 at night.

Behind all these systems is something far more crucial than fans and sump pumps. They all depend on electrical power and flowing water — or rather, the heat exchange made possible by flowing water. And all of that depends on the Energy Center, a two-story building made up of a series of opaque vault structures similar to those glassed vaults that enclose the agricultural unit of Biosphere 2. It sits against the ridge immediately to the west of the IAB with its three huge concrete cooling towers a short distance away on its north side.

With a complete power failure in summer, the greenhouse effect could quickly shoot the temperature inside to uninhabitable extremes. If it's one hundred degrees outside, in less than an hour it could go up to one hundred and thirty, forty, or even one hundred and fifty degrees inside. Plants would die en masse. The heat would force the Biosphere 2 crew to break closure and get the animals and themselves out as fast as possible. Such a scenario must be prepared for. To prevent a power loss at all costs, SBV supplied the Energy Center with three separate generators. The total capacity is about 5.5 megawatts, but Biosphere 2 could probably operate adequately on only a third of that. The local town of Oracle's power grid, although vulnerable to outages (especially during the summer lightning season), provides another source. It would take four separate and simultaneous failures — all three generators plus the local power utility — for Biosphere 2 to completely lose its electricity.

The Energy Center also produces the hot and cold water for heating and cooling, which is accomplished by piping of hot and cold water into Biosphere 2 to the airhandlers, which would cycle air as needed to cool or warm it. Those waters and the inside of the pipes carrying them are actually "outside" of Biosphere 2, but permit exchanges of heat — playing a part in Biosphere 2's energetic openness. It would take a serious earthquake — rare in Arizona, but not unheard of — or a backhoe accidentally splitting two sets of pipes to cut off the two flows of water. Dempster tried to prepare for everything he could: Biosphere 2 was designed to withstand up to an inch of terrestrial movement.

"I'm very skeptical about computer systems controlling everything," Van Thillo says between bites of a sandwich at lunch in the dining hall in May of 1990. He has already spent most of his day in the long process of shaking down airhandlers, irrigation systems, pumps, motors, and valves, so the idea of depending on absentee captains outside in Mission Control doesn't thrill him. "I haven't seen one that works absolutely perfectly yet."

If Biosphere 2's computer system were to fail, the controls of mechanical apparati will automatically adjust themselves to a preset level, such as a thermostatic control set to turn on airhandlers if the temperature exceeds seventy degrees. If the automatic mechanisms fail, they are backed up by manual controls. Making adjustments to keep up with changes in temperature might run the biospherians ragged, but this might nonetheless be crucial to the project's survival in the case of an emergency.

The build-up of an unexpected trace gas also could force the breaking of closure. Or an uncontrollable plague-like insect outbreak in the farm might leave the biospherians with little food. One of Linda Leigh's nightmares was that one of the crew might accidentally leave a hose running overnight, flooding the desert. The desert has a salt playa, an area where water repeatedly collects and evaporates, leaving behind salt deposits. An accidental flooding could float the salt elsewhere, possibly carrying lethal doses to other areas. So timers that shut off hoses automatically were installed to reduce the likelihood of this particular nightmare.

Intelligent Systems

The cornerstone rule of Biosphere 2 is that *no operation of the nerve system shall harm life*. This can be contrasted to Asimov's First Law of Robotics: no harm to humans. This is buttressed by rules which govern the minimum and maximum limits of temperature, humidity, water quality, etc., as determined by the ecologists to be necessary for the life systems. A subsidiary rule such as "all equipment shall be tuned and run in the most energy-efficient manner" cannot overrule the top-line rule of the preservation and enhancement of living organisms.

One of the rare aspects of the nerve system approach being taken is the dual use of human intelligence and artificial intelligence *at all levels*. Automated sensors connected to computers and people trained in naturalist observer science provide two types of 'hardware' working at different speeds, with different programs or 'software' to interact with each other.

Artificial intelligence — the attempt to develop computers that can truly learn and think, not simply process data in a programmed fashion —has made great advances. Expert systems, developed for specific applications such as medical diagnosis or oil exploration, have proved extremely effective — in some cases, more reliable than human experts themselves. Computer-based systems for control of complex industrial operations have also been achieved, using real-time data analysis and response to maintain quality and production levels.

The nerve system of Biosphere 2 required, however, a new paradigm for artificial intelligence systems. Complex data acquisition/computer/ managerial systems had never been applied before to an ecological reality where there are dynamic plant and animal populations, critical recycling air and water systems as well as a technical system. It took some time before the engineers realized that observations of indicator species — plant or animal — which were sensitive to changes in nutrients, acidity of soil, or water quality could be as accurate a sensor as an electronic device. Natural and electronic sensors working together would offer far greater sophistication and capability than either alone. The new paradigm required that artificial intelligence meet and synergize with the natural intelligence of both the humans and of the ecological systems.

The dual-brained system was made to operate at five levels. At every one of these levels, the humans are able to intervene. Nothing is entirely left to artificial intelligence, which is being constantly updated and improved as the system-as-a-whole *learns*.

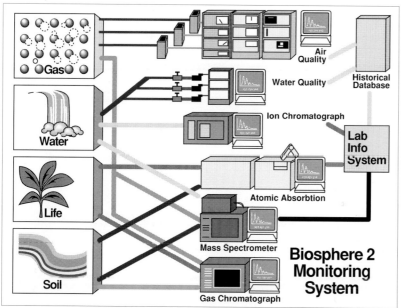

Gas

Water

Life

Soil

Air Quality

Water Quality

Historical Database

Ion Chromatograph

Lab Info System

Atomic Absorbtion

Mass Spectrometer

Biosphere 2 Monitoring System

Gas Chromatograph

Left: Graphical representation of the monitoring system. Humans interface at all levels of this system.

Below: The Test Module link with Abigail Alling during her five day stay shows one of the several ways in which biospherians will stay in contact with Biosphere 1.

At the first level, there is contact with the immediate physical world, either through sensor or human observation (the "naturalist's trance"). A sensor may show some change on a screen, but the biospherians can actually go there and take a look for themselves. This is also the level of actuators: if the flow of water is too much, it is automatically reduced. There is a system of "rules": if so-and-so happens, then do such-and-such.

At the second level, the data coming from sensors and observations is filtered and packaged. Filters cut out the "noise" in the system. Data is packaged for easy handling and building into patterns. So, for example, one can find out the range of temperatures of a region day by day, or the variation in average temperature week by week, season by season.

At the third level, that of information processing and networking, it is possible to see the working of a whole biome; indeed, to see whether it is working. Here, there is a supervisory function that involves the scrutiny of specialists and advisors. It is possible to review the operation of the various controls and rewrite the rules based on experience.

At the fourth level, there is the construction of the "story" of the functioning of Biosphere 2. Here is the archive of events, the meaningful narrative one might say. The perspective is "global" and includes the effects of the various biomes upon each other.

At the fifth and highest level, interbiospheric communication is taking place. This is called the level of *understanding*. The Biosphere 2 nerve system is networked across the Earth, linked through the most advanced fiber optics to other centers of biospheric knowledge.

The potential for extending this type of approach to our global biosphere is a vision many thinkers in past times have seen: from Fuller's *Operating Manual for Spaceship Earth*, to Vernadsky's concept of noosphere. It is now at the beginning of both the space and information ages that the tools are at hand to assist humans in the wise stewardship of Earth. Humankind and the biosphere of Earth share a common destiny, and intelligence will be needed to help us achieve it together.

10
The Human Factor

"It was a risk. But man often takes risks when he tries to penetrate the unknown. Didn't the great explorers risk their own lives in their travels to distant lands? And weren't the men who explored the poles at risk? And those that scaled towering summits? What about the physicians that inoculated themselves with the lethal bacilli of plague, yellow fever, cholera in order to learn about the onset of these diseases, how to prevent and treat them?"

Vitaly Volovich

Numbers

Crew size was obviously a critical decision that needed to be made early on. The Institute of Ecotechnics had experimented with small task group sizes for a number of projects throughout its history. As a prototype for a Mars colony, Biosphere 2 was its most multi-faceted challenge to date. Eight people appeared sufficient to handle the workload necessary for such a venture. Eight were numerous enough to keep each other company. Other researchers had come to similar conclusions about the ideal size of a small group. Hyorishi Kuriki of Japan's Institute of Space and Aeronautical Sciences recommended eight as an ideal number for a lunar base or space station. Eight was the original number for the crew of Space Station *Freedom*. The number eight had also been suggested as ideal for an international manned mission to Mars. And eight had long been the size of the basic unit of military life; a squad consists of eight soldiers.

Training

When SBV management looked at all the requirements for potential "biospherians" — those who would operate, live in and develop a complex highly integrated system such as Biosphere 2 — they seemed staggering indeed.

Biospherians must be "ready, willing, and able" to fulfill at a very high level, "rapidly, perfectly, and easily" the following roles:

- Naturalist observer
- Ecological systems analyst and synthesist
- Gas chromatographer
- Advanced computer programs user
- Network communicator
- Skilled farmer-gardener
- Chef level cook
- Mechanic and preventive maintenance quality controller
- Speaker and writer
- Manager of two complex areas, one of which is a biome or atmosphere and the other a technical system
- Researcher

Fourteen biospherian candidates emerged and all were in training before several years for the first closure experiment was to begin.

CHRISTINE HANDTE

On expedition to Antarctica, a biospherian candidate scans the water for ice.

Opposite: Biospherian training involves learning complete deep diving and blue ocean sailing skills aboard the R/V Heraclitus. Some of these, like turning the capstan to raise the mainsail in this photograph, involve coordinated group skills as well.

The R/V Heraclitus *at rest in the Antarctic waters.*

ISOLDE DROSCH

Two biospherian candidates preparing to explore the underside of a neighboring iceberg during the R/V Heraclitus' *expedition to Antarctica.*

A diving program aboard the *R/V Heraclitus* is one of the major training devices. Like NASA and the Soviet space program, SBV considers it one of the best methods available to develop the right stuff in its crew. Diving introduces candidates into a new dimension of gravity, as well as the challenging disorientations of an unfamiliar world. Part of the program's usefulness is also the exploratory nature of the ship itself; in its fifteen year history it has undertaken a variety of ambitious expeditions. In the late 1970s it explored the Indo-China Sea and the Indian Ocean; in 1980 and 1981 it sailed up the Amazon on an eighteen month ethnobotanical expedition; from 1983 to 1985 it completed a tropical circumnavigation of the globe; and it undertook a circumnavigation of South America in 1988 and 1989 which included a austral summer voyage to Antarctica. Its most recent mission has been to collect coral reef species in the Caribbean for transport to Biosphere 2.

ISOLDE DROSCH

The training periods aboard the *Heraclitus* did more than generate an esprit de corps combined with the essential nuts-and-bolts practicality of learning to operate in "self-contained underwater breathing apparatus" (SCUBA) systems. Its scientific missions brought in a wealth of specimens and samples from ocean and shore. In addition, the rigors of operating an eighty-two foot sailing vessel on the high seas combined physical fitness and experience of the discipline of a shipboard command system: a captain with ultimate authority, a first mate, and designated crew members, each with their appointed tasks and areas of responsibility.

The mobility of its training laboratory has also allowed SBV to emphasize its "spaceship Earth/global village" orientation by putting its personnel in direct contact with diverse cultures and races. Many of the candidates, such as Sally Silverstone and Kathleen Dyhr, were already well-traveled world citizens intimately acquainted with the cultures and challenges of distant lands. Although more an auxiliary rather than a specific prerequisite for biospherian candidacy, familiarity with and active interest in the global community has certainly been a relevant consideration.

SBV also has an international mix in its staff. The fourteen candidates come from six countries: the U.S.A., the United Kingdom, Australia, Mexico, Belgium, and Germany. "To be wise in the ways of the world," meaning an awareness of the status of the world as a whole, was one of the goals John Allen set for the crew.

By 1989 all the candidates fulfilled the requirements which had originally been listed. All team members had a sound knowledge of the life sciences in general, and ecology in particular. What Augustine had to evaluate was more the balance of their qualities and skills: their leadership abilities, their medical and veterinary knowledge, their sophisticated technical know-how. She also considered the long-term personal goals of the candidates. Some of the candidates, such as Van Thillo and Zabel, had come to the project in its earliest stages with the goal of being a member of the first team to go in. She kept this seniority in mind. A key component in the final selection was a kind of emotional mobility in the candidates: an ability to "shift roles abruptly," as Allen puts it, to accept any position in the group hierarchy and do any job as needed. In any case, the initial team to enter Biosphere 2 would be only the first of many. A second team was already in training. The six biospherian candidates working on the outside during the first two-year closure would most likely be a part of the crew for the next experiment.

On September 12, 1990, a press conference was held announcing the eight biospherian crew members for the first two year experiment. The international media spread the news: Bernd Zabel, 41, from Germany, will serve as the Crew Captain; Sally Silverstone, 35, from England, as Co-Captain and Information Systems Director; Linda Leigh, 38, from the United States, will be Director of Biosphere 2 Research and of the terrestrial wilderness biomes; Abigail Alling, 30, from the United States, Director of Biosphere 2 Development and of the marine biomes; Roy Walford, M.D., 66, from the United States will be the Medical Officer; Taber MacCallum, 26, will serve as Analytical Systems Manager; Jane Poynter, 28, from England, will be Manager of Intensive Agriculture Systems; and Mark Van Thillo, 29, from Belgium, the Technical Systems Manager for Biosphere 2.

129

Keeping Healthy

Dan Levinson, M.D., led a team of physicians at the University of Arizona Medical Center that evaluated the physical condition of the candidates. Augustine kept her hands off that part of the evaluation process. "Like any business," she says, "the medical records of our personnel are strictly confidential. We were appraised of their good health on a comprehensive level. Are they physically competent for the task or not?" Some specific recommendations were made as a result of those examinations, however. Before Zabel goes in, his wisdom teeth have to come out. Augustine laughs about getting everyone off to the dentist being one of her hardest tasks. Before closure, the crew will undergo more extensive testing and will continue to be tested during the experiment to obtain medical data on the effects of the closed system environment.

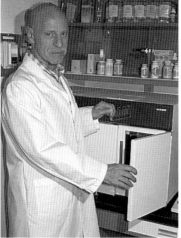

C. ALLAN MORGAN

Biospherian Roy Walford, M.D. of the pathology department of UCLA, preparing the medical facilities for Biosphere 2

Augustine has been approached by numerous individuals interested in conducting sociological, anthropological, or psychological studies of the crew during its long isolation. Her response has been that the crew's private lives are private. If they want to participate in some study on their own, they are free to do so, but no such study is being undertaken with SBV's cooperation or sponsorship. The management wants nothing to interfere with the managing and studying of Biosphere 2. Still, the hottest topics of interest for the news media, and a subject for amused conjecture for Biosphere 2 watchers, are the love lives of the crew once the doors have closed. Four men, four women. None of them married to one another. Scientific inquiry may be their primary objective here, but it's hard to imagine that eight healthy adults will put romance and sex on hold for the entire two years. Curt Suplee of the *Washington Post* put it thus: "Will there be sex in the Biosphere? Of course, but who cares ... those bidding for a berth in Biosphere 2 are in it for the love of the idea, not the colleague down the hall."

Most of the biospherians seem to think that something quite irrevocable will change in their lives, culturally as well as intellectually and emotionally. As Poynter says, only time will tell just what those changes will be. The private journals of the crew may someday reveal some interesting aspects of the dynamics of isolated colonies that will be applicable in the new world of space exploration. Of course, in this instance, the isolation is only partial. Radio, television, and video communication with colleagues, families, and the outside world will occur at an intense level. The eyes of the world will be upon them, and the biospherians know it. Rather than feeling isolated, they may end up feeling like goldfish in a goldfish bowl. But while SBV may try to shield their private lives from the public gaze, their physiological conditions will be under constant medical scrutiny, under the supervision of biospherian and medical doctor Roy Walford.

Walford points out that Biosphere 2 is not a space station or extra-planetary colony, either of which would require that almost all medical needs be handled on the spot. In off-Earth conditions, the amount of sophisticated training required of the diagnostic and therapeutic personnel, as well as the expense of the facility, goes up astronomically. The medical facility in NASA's planned space station is probably going to cost from thirty to fifty million dollars. NASA will have to build miniature versions of all their laboratory equipment. Biosphere 2 can use the regular equipment because it doesn't have the same space limitations. Since the crew is small, the biospherians will be able to do some medical analysis by hand, such as the bacteriology and blood work, without the high-dollar equipment used to speed up procedures in large facilities. Walford has six to eight different culture media available for specimen testing and a system for identifying fungi and molds.

While Alling and Leigh were in the Test Module, a borderline drop in their blood potassium was noted. Initially it was thought this could be due to unconscious hyperventilation from emotional excitement. But it turned out that it was due to under-eating — from excitement — and having a lot of potassium-rich sweet potatoes in their diet. Aside from that, there was little in those early experiments to alert them to probable complications in Biosphere 2. Walford considers toxicity the most worrisome unknown.

C. ALLAN MORGAN

Linda Leigh in Test Module taking samples of her own blood during the twenty-one day "Vertebrate Z Experiment".

In addition to Dan Levinson, Don Paglia, M.D., also a medical professor at UCLA, and other physicians with various specialties will consult through computer, telephone, and video. Mission Control will be able to transmit X-rays, electrocardiographic tracings, and other diagnostic information. Levinson hopes that everyone inside will not only stay free of illness, but actually get healthier by being in a pollution-free environment. "This is going to be a marvelous test, to compare a pristine environment with this mess we have, called Biosphere 1."

In the early planning stages, SBV asked Walford to find out if antibiotics and other essential drugs could be made inside. He ascertained that the manufacture would be too complicated for the facilities. All medicines have be taken inside with them.

Music is a common interest of several of the biospherians. According to the ancient philosophers, music may serve the health of mind and body together. Leigh will be taking in some of her woodwind instruments and perhaps the soprano sax that she's been meaning to learn. Poynter plays the flute and the balalaika. MacCallum will be taking in his cello, synthesizer, and kanga drums. Van Thillo does not play an instrument himself but wants to take in as much of his music library as he can. His tastes run from Ravel to Ornette Coleman to folk music. Another common interest seems to be science fiction novels, along with other personal reading favorites like the complete *1001 Arabian Nights* for Silverstone and *Swamp Thing*, a British comic book with an ecological theme, devoured regularly by Leigh. Zabel wants to catch up on Plutarch's *Lives* and Nietzsche; he hopes to learn Russian over the two years and to find time to do some painting. They're all taking in books related to their particular interests: for Leigh, plant pathology, for MacCallum, chemistry, for Alling, marine biology.

Eating Well

One of the highlights of Biosphere 2 by just about anyone's standards is the food. SBV is not kidding when it lists as a requirement for its biospherians, "chef-level cook." Food and its preparation is taken very seriously, and it's not just a well-balanced diet they're providing — it's *cuisine*. The best organically grown, chemical-free ingredients will spill out of their plots, through master gardener Jane Poynter's processing area in the basement, and into a fully equipped kitchen that would be the envy of many a fine amateur chef. The processing area is stocked with threshers for the grains, mills for grinding everything from wheat to coffee, presses for extracting peanut oil and sugar, coffee and peanut roasters, a rice huller, seed cleaners, drying ovens, and storage bins. The kitchen offers a conventional oven and range (electric), a broiler arrangement with the heating element above to reduce smoke, a microwave, a cuisinart, a coffee grinder, and the usual assortment of gadgets. No hibachis or barbecue grills, however, as this is strictly a no smoking area. If a careless cook does burn the toast, the soil bed reactors will have to deal with the accident. How much of that they can absorb is all a part of the experiment.

Jane Poynter planned the preparation of the food along with the preparation of the agricultural area. She is quite pleased with the results of the years of work. "We have all the ingredients in the tropical garden for very fine cuisine, from French herbs to tropical spices: rosemary, thyme, coriander, parsleys, lemon grass, sage, garlics, onions, cloves, nutmeg, sesame, ginger, horseradish, peppers, and chiles."

There are two cacoa trees in the rainforest to tap for a small supply of chocolate. Poynter allowed that this would be only for a treat, like Christmas cookies, as two young trees just won't produce very much. A good selection of dwarf coffee trees, already producing good crops of beans, were arranged in a shady row in front of the habitat. For flour they'll have rice and several types of wheat, including both pasta and bread wheats. There's a machine to process the pasta, too.

A Biospherian Feast

Artichokes with grilled goat
    cheese and chopped parsley
Peanut soup with chili
Tilapia stuffed with kumquats
    and wrapped in lemon grass

Rice with mushrooms
    and herbs
Ratatouille Strawberry tarts
    and lemon sherbet
Café au lait

A cornucopia from the intensive agriculture biome.

KENT WOOD

Daily Life

The daily lives of the biospherians will obviously be a little different from their lives in Biosphere 1, but they'll probably soon fall into repeated routines like most creatures do. The main difference will no doubt be the scrutiny they will have to learn to live with. The doctors will be monitoring their every ache, their every stumble, their every calorie of food. Reporters will be watching for every puzzle or problem that may signal a crisis looming on the horizon. There will even be video cameras installed around the wilderness biomes and the agricultural biome that can be turned on and off by either the biospherians or Mission Control. Arrangements are in the works to offer some access to the television cameras for broadcasting to the public. There will also be walkways all around the outside of Biosphere 2 to bring thousands of visitors right up to the glass walls.

There is much that is undeniably futuristic about their daily lives here. But there are also reminders of older ways of living. Grinding their own flour and pressing out their own cooking oil are only a few examples of old practices revived. On the other hand there are some inconveniences that are not so picturesque. For one thing, the women cannot bring in a two year supply of tampons. It's not a matter of storing the items; it's a matter of disposing them: tampons are not biodegradable. So they're using natural sponges supplemented by washable pads that they wash and reuse, just as women have done for thousands of years. Linda Leigh used natural sponges quite successfully when she menstruated during the three week experiment in the Test Module.

Another modern convenience that will be left behind is toilet paper. The waste disposal system will not be able to handle the problems of decomposing the quantity of toilet paper that eight people would produce. They'll simply wash off in the traditional European style. Long, leisurely showers will also be unavailable; they'll have to use less water than they may have been used to, though a higher water pressure will almost reproduce the sensation of a full shower. The ration of water per person per day is determined by the capacity of the elegantly simple system of water treatment worked out by architect Hawes and installed down in the basement.

In the southwest corner of the lower level, beneath the farm habitat, three small tanks lie in a row next to a bank of windows. Two of the tanks measure eight by sixteen feet, the third eight by eight. The first tank contains layers of various types of gravel substrate topped by soil and marsh plants — tall grasses, cattails, bulrushes, and reeds. Similar aquatic plants grow in the second tank where water is exposed in open meandering channels. The smaller third tank contains much more open water with small islands. These three tanks, totalling just over 300 square feet, comprise the entire treatment system for all the wastes produced by the human occupants. Here, through nature's alchemy, green plants and microbes combine with sunlight to make dirty water come clean again.

Wastewater from showers, sinks, laundry tubs, and toilets flows into the gravel of the first tank where microbial and filtering action breaks down solids. The water then flows into the second tank where more plants and microbes work together to break wastes down further and use them up as nutrients. In the third tank, more of a pond than a marsh, the water receives a final level of biologic treatment before it is pumped into a utility water tank where it is stored and later used to irrigate agricultural crops. Hawes' marsh system is designed to handle about six hundred gallons of wastewater each day, which means that the daily water allowance for each biospherian for all kitchen, bathroom, and laundry needs is a maximum of seventy-five gallons. Since all showers, sink taps, and toilets are designed to economize water use, the daily allotment should be more than adequate for all human needs, with plenty left over for things like cooking and cleaning.

C. ALLAN MORGAN

Jane Poynter wields a hoe in the intensive agriculture biome.

All the residents are expected to put a couple hours of work into some part of the agricultural area every morning. This includes the planting, sowing, irrigating, weeding, fertilizing, transplanting, mulching, harvesting, pest control, and composting of the crops, as well as manual pollination of a few of the crops to augment the natural services of the insects; the feeding, doctoring, breeding, midwifing, and culling of the domestic animals; the raising and harvesting of the fish in the rice and azolla system, and soil monitoring and treatment.

They must also monitor a mechanical system on the scale of the innards of a cruiser. Every day the "First Mate" and an assistant must check a list of motors, pipes, gauges, levers, pumps, switchboxes, vacuum blowers, hoses, fans, cables, ducts, and dials that runs to several pages. There are approximately fifty miles of pipes, over two hundred motors, one hundred and five pumps, and about sixty fans; so far no one has calculated the mileage of cable and hose. Every ten days each of the sixty algae scrubbers must be scraped clean by hand. The sensing equipment monitoring air pressure, oxygen and carbon dioxide levels (along with the levels of other gases), temperature, and humidity must be checked several times a day. Each biospherian has a roster of duties to attend to involving their own expertise such as Walford in the medical lab, Poynter in the food processing, or Zabel in maintenance of the structure — no small task in a building this size.

Observations will be made in the wilderness biomes every day as well. Each biospherian is responsible for keeping a detailed record of plant conditions along a designated transect every week. These particular transect studies will provide a complete two year life history of numerous species. Observations in the wilderness will tell them when it should rain and when to lower the temperature. Observations will also help let them know when to alter the tides. The biospherians have to make sure the pH of the ocean, the marshes, and the stream stays within acceptable ranges to avoid any inadvertent, unnecessary extinction. Some of the wild animals must be caught, examined, and weighed periodically. The hummingbird feeders must be replenished regularly, as well as supplemental food processed and made available to other inhabitants

Bernd Zabel preparing the rice paddies in Biosphere 2.

C. ALLAN MORGAN

of the wilderness biomes, such as the galagos, until it is certain that the system has reached adequate production rates.

The daily cycle of activities is not simply a sequence of chores; it is a way for the biospherians to manage themselves. At breakfast time, they meet to take a look at the day ahead and to review the state of 'their world'. Everyone is engaged in the care of both the habitat and the agriculture, as well as the Biosphere as a whole. The old saying, "No work, no eat!" is emphatically said here. But the biospherians must also be finding some greater value than mere survival through their efforts and experiences. What new horizons will they come to see during their sojourn? How will they communicate to others their discoveries at this meeting point of life, technology, and humanity?

Bottom left: Taber MacCallum making a parts per billion nitrate analysis.

Below: Abigail Alling checking out the ocean floor.

KENT WOOD

PETER MENZEL

11
The New Explorers

"The people, yes, the people ... and overhead a shovelful of stars ..."

Carl Sandburg

Abigail Alling

The shores of the Atlantic Ocean were never far away in the childhood Gaie spent between her family homes in Maine and on an island off the coast of Georgia. Her world was mainly ocean, inhabited by seals, dolphins, and whales. Later on in life, she would remark that: "We are all sea people and the planet — as Arthur C. Clarke pointed out — would be more properly called planet Ocean than planet Earth".

Gaie earned her Bachelor's degree in biology from Middlebury College in Vermont, and her Masters in ecology, with a focus on marine mammals, from Yale University, School of Environmental Studies. By her mid-twenties, she had published several papers on dolphin and whale research. She has studied at both Cambridge University and Harvard, and received awards and fellowships, including one from the Watson Foundation.

Her research with marine mammals included work in Greenland, Canada, British Colombia, South America, Antarctica, Asia, China, and the Caribbean. For five years she lived in Sri Lanka, studying sperm whales for a World Wildlife Fund Project. She also worked with Roger Payne to set up a marine mammal protection unit for the United Nations and the government of Sri Lanka.

In 1986, she followed through with her marine work and became the Project Director for the *R/V Heraclitus*, proposing improvements for more extensive ocean research. Colleagues from ORCA (Oceanic Research Communication Alliance) planned the release of two captive dolphins — Joe and Rosie — back into the wild, which she undertook and successfully accomplished off the coast of Georgia. Subsequently, Gaie organized and led an expedition of the *Heraclitus* around South America, including a voyage into the waters of Antarctica to study and track the summer residents of a humpback whale population. In her expedition log, she says: "We are whalers of a modern time — trying to bridge this past with the future, to try and find out how much time we have left, how many whales were left, would we be able to do anything or was it already too late."

It was a short step from such marine research to coordinating the design, collections and implementation of the ocean and marsh biomes with Walter Adey. Gaie had made herself a biospherian, one of the new explorers. As she wrote; "The hint of essence where qualities are rare, the land untouched, this is what the explorer seeks to know."

JEFF TOPPING

Abigail Alling — born New York City, USA, 1959.

Opposite: The eight biospherians: four men, four women; four nationalities; eight individuals of extraordinarily diverse adventures and accomplishments. From left: Mark van Thillo, Taber MacCallum (top), Abigail Alling (bottom), Jane Poynter, Linda Leigh, Bernd Zabel (top), Sally Silverstone (bottom), and Roy Walford.

C. ALLAN MORGAN

Linda Leigh — born Racine,
Wisconsin, USA, 1951.

Linda Leigh

From the tundra of Alaska, elegant and spare, to the jungles of Venezuela where life tangles and tumbles over life, Linda has canvassed the spectrum of the biosphere. Her mother's love of wildflowers introduced her to the beauty of nature. She was inspired by a teacher who had a great love of biology and began to be interested in the theory and science which underlay the natural world. Later, at the University of Wisconsin at Madison, she turned to the study of botany and trained as a naturalist.

She was team botanist for a National Science Foundation project to replant native grasses in the midwestern prairies. She finished her Bachelor degree in Field Botany and Ecology at Evergreen State College, Olympia, Washington where she studied elk populations and assessed the possible effects of reintroducing the grey wolf. This was no armchair study. She had to think and sense like a wolf! Moving further north, her next project was in Alaska, working in an interdisciplinary team on a management plan for the whole peninsula. She recalls her close encounter with a nine foot bear: "I suddenly remembered what an old - timer had told me — when they charge make yourself seem as big as you can." She followed the advice and survived to tell the tale!

From Alaska she was called to the desert regions of Arizona and Mexico and, in 1979, work with ethnobotanist Richard Felger brought her to Tucson. It was inevitable that she would link up with Space Biosphere Ventures when they began operations in the area. When she learned that a position might be open for her on the design team for the biomes, her response was emphatic: "Yes!"

In 1987 she travelled to London, England to talk at the conference on closed ecological systems held at the Royal Society. It was a great moment for her was when the library historian of the Society turned the pages of the roster of members to show the signatures of Newton and Darwin. She recalls: "I thought not only of the human giants upon whose shoulders we stand now in the twentieth century in our attempts to understand the biosphere, but also of the biosphere and all the giants — great and small — who choreograph the marvelous dance of life on Earth. I thought of the great debt we have accumulated to these many unsung heroes of the biosphere and the hope that Biosphere 2 may enable me to repay some part of that debt."

Taber MacCallum

Taber MacCallum — born Albuquerque, New Mexico, USA, 1964.

At the age of five, Taber's father, an astrophysicist at Sandia National Laboratory, took him to see the lunar landing site through a telescope. Space travel became central to his life and the dream was fuelled by the science fiction of Ray Bradbury and the space hero Buck Rogers. However, his explorations began on Earth. Adventures took him to the south, west and east of Europe, on the trans-Siberian railroad and into Japan, China and Tibet. While in the Far East, he chanced on the opportunity to work aboard the ocean-going sailing ship, the *R/V Heraclitus;* and there he came to learn of Biosphere 2.

The *Heraclitus* was engaged in collecting microbial samples for Clair Folsome. Taber was fascinated with all aspects of seamanship and became the diving master, training several biospherian candidates. Coral reef ecology became his focus of interest. He travelled with the *Heraclitus* over four oceans, visiting forty ports in twenty-five countries, a journey of thirty thousand miles. His experiences at sea led him to propose a corollary to Murphy's law: "Bad things always get worse at night!"

Taber applied for a position with the Biosphere 2 project at the end of the expedition. He took on the development and management of analytical systems. Building on his experience as a diver, he first worked with staff at the Environmental Research Laboratory and then directly with the Hewlett-Packard team contracted as consultants. He was responsible for the design of the analytics lab that is incorporated inside Biosphere 2, a design that uses no toxic or non-renewable chemicals, a feat on the cutting edge of ecological technics. The human medical program became his second area of expertise.

Taber is a graduate of the International Space University's initial session at MIT in 1988. His diverse skills and leadership abilities placed him as team leader in the practicum course which served as the final exam. Their task: design a lunar base.

He will be twenty-six when he goes inside Biosphere 2 and may well live to walk on Mars. *"Ad astra"* says the young researcher. "To the stars!"

C. ALLAN MORGAN

Jane Poynter — born Surrey, England, 1962.

Jane Poynter

Jane had read Isaac Asimov and Arthur C. Clarke since she was eight or nine but never suspected that she would be taking part in an historic venture leading into space. Born in England in 1962, she graduated from St Michael's School for Girls in Sussex, took a business course in London and headed, at the age of eighteen, "down under" to Australia. It was there, at one of the Institute of Ecotechnics' consultancies where she was mastering the art of mustering cattle and of regenerating desertified pastureland, that she heard of the Biosphere 2 project from Mark Nelson.

A participant in the biospherian program since 1985, Jane's training included two six month voyages on the *Heraclitus*. Her first took her from Sri Lanka to Egypt. She learned diving and explored some of the richest coral reefs in the world. She also came face to face with a white tipped shark in a breeding ground in the Red Sea! Her next trip involved her, along with Abigail Alling, in the first successful release of previously captured and trained dolphins back to the wild.

At Space Biosphere Ventures, Jane's responsibilities included the Insectary for the raising of the forty-two species essential for Biosphere 2, some of which have never been raised before. She worked with Randy Morgan, head of the Cincinnati Zoo's insectary, and went to Hawaii to work with Scott Miller of the Biship Museum, SBV's chief entomology consultant.

She managed the Tropical Garden prototype greenhouses where cropping and animal systems were developed for the Intensive Agriculture Biome of Biosphere 2.

The challenge of being involved in the first two-year closure kindles her infectious enthusiasm: "I wouldn't want to be anywhere else. I know the other biospherians well, some from field expeditions, others from work here in Arizona. I know what they are like in a tight situation and I know I can count on them. I've always tried to live an interesting and diverse life and it's going to be that inside Biosphere 2."

"I sing. In Biosphere 2, I'm going to sing to the rainforest and desert...One day I would love to sing *Songs of Distant Biospheres* to complement Arthur C. Clarke's *Songs of Distant Earth*."

Sally Silverstone

Raised in London, England, Sally claims she learned there nothing of the way grass grows but everything about the life of the city's streets. At the age of eighteen, she packed her bags to take her first job which was in a home for abandoned children in a village in Kenya. She was out in the bush with one hundred children and had to learn fast. Sally became intrigued with the people and cultures of East Africa and fascinated by the wildlife and plants of the exotic terrain.

After a year in Kenya, she returned to England to study social work at a college in Sheffield and received a degree four years later. During this time, she met John Druitt, who was later to become director of a tropical rainforest project in Puerto Rico consulted by the Institute of Ecotechnics. However, tropical projects were far from their minds. With some friends, they tried their hands at farming some poor land near Sheffield through which Sally gained experience of management and agriculture.

After graduating, her sights were set again on distant horizons. Initially hired to work at a children's home near Calcutta, India, Sally was suddenly transferred to another project near Bihar when the manager of the Calcutta project took off one day and left her with a houseful of retarded children. In Bihar, she learned "quite a bit" about survival. She also had her first taste of what ecological disaster looks like. "Frankly," she says, "I was more than a little desperate about what could be done about it."

Back in England after three years in India, she had news of her old friend John Druitt, now in Puerto Rico. He was replanting in deforested areas, interspersing economically valuable trees such as mahogany and teak while preserving the integrity of the rainforest. She went to work with him — and came to hear of the Biosphere 2 project.

When she joined Space Biosphere Ventures as a biospherian candidate, Sally took on the tropical agriculture greenhouse. Her managerial capacity, however, soon moved her into the complex task of making sure that everyone was on track for the critical path to the birth of Biosphere 2.

Distant shores still call to her. "When I heard of Biosphere 2, my first vision was a biosphere on the face of Mars, blooming. The only way to explore space is going to be in a biosphere. I mean, who wants to stick with just one planet?"

C. ALLAN MORGAN

Sally Silverstone — born London, England, 1955.

C. ALLAN MORGAN

*Mark Van Thillo — born
Antwerp, Belgium, 1961.*

Mark Van Thillo

Raised just outside the city of Antwerp in Belgium, Mark comes from an old and prominent Flemish family. His father was a mechanical engineer. As a child, he amused himself by wiring up appliances discarded in his affluent neighborhood.

He graduated from Don Bosco Technical Institute in Antwerp and made the first of his exploratory expeditions, which was to India and Asia for six months, travelling about on foot. On his return to Belgium, he worked as a production assistant in a refining plant where he acquired an appreciation for complex industrial facilities. Then, restless to see more of the world, he embarked on a private ecological expedition for two years in Central America, in which he was responsible for the outfitting, maintenance and repair of transport and other equipment on the journey. His experiences in Central America deepened his fascination with the complex 'machinery' of the biosphere. His interest in space also began to deepen at this time. An avid reader of science fiction, he began to read the scientific literature on space exploration and, in particular, the astrogeology of Mars.

The opportunity arose of collaborating with Silke Schneider, who would later also become a biospherian candidate, in a personal exploration of the American frontier. "On horseback, and with an old truck converted to serve as covered chuck wagon, we traversed the 1,200 miles of the Santa Fe Trail and Chisolm Trail." He met people from the Ecoworld project originally conceived by the Institute of Ecotechnics who were working in the high semi-arid grasslands south of Santa Fe. There he began to study life systems as he had previously studied mechanical ones.

Mark applied for and was accepted into the biospherian training program in 1984. His first training endeavor was aboard *R/V Heraclitus*. As he says, "A boat is the nearest thing to a closed system". He rose to the position of Chief Engineer and achieved a certificate as an advanced open water diver.

On the Biosphere 2 site, he was in charge of quality control, installation and maintenance of all mechanical support systems. With his string-pole 6'1" frame, he was a popular sight as he zoomed around in his red Ford pickup checking on the crews.

"If it really works, then we have an invaluable tool for space habitation. Next challenges would probably be Antarctica and then underwater biospheres. Ecology is very important to me, but you see, we are travelling in space already — this planet is our first spacecraft."

Roy Walford

Roy, now aged 66, is a man with more than one life. His scientific specialty is formally called "gerontology", the study of aging; but, for him what matters is the art of staying vital. Interest, humor and diversity are crucial for the quality and longevity of life.

As an undergraduate, he studied at CalTech, and then he graduated from medical school at the University of Chicago. He balanced this academic life by working out a system with a friend to beat the roulette game in Reno, using their thirty-five thousand dollar winnings to buy a sailboat and cruise the beautiful Caribbean seas for a year. Roy's second life is for living!

However, his career in research is nothing short of illustrious — with three hundred published scientific papers, six books and many prizes in his field. Roy's research work in the field of molecular genetics deals with the effects on aging of transferring genes from one species to another. He is on the cutting edge of modern experimental biology. In addition to his scholarly scientific books, *Maximum Life Span* and *The 120-year Diet* thrust him into the public arena as an educator on the most current theories of how to extend the healthy, productive years of life.

He has ventured far and wide in his search for knowledge: to the exotic lands of South America in search of the most short-lived fish; and to the subcontinent of India in search of yogi masters capable of lowering their body temperature. In his second life as an adventurer, Roy includes being a poet and journalist. In 1968, he met Julian Beck of the Living Theater in Paris, which led to Roy starting a street theater group in Los Angeles. Through theater work, he met Kathelin Hoffman, member of the Board of Directors of the Biosphere 2 project.

Life number three, as a biospherian, began about four years ago. One of his major contributions has been a computer program that takes data on what one has eaten, translates this into calories and nutrients, and tells one what to eat for the remainder of the day to fulfill nutritional requirements. Inside Biosphere 2, he will be responsible for the medical monitoring program and communications and data aspects of the computer system.

"Biosphere 2 brings my first and second lives together. The next phase of hard-core biology when molecular genetics has matured, will be Systems Analysis which has needed the development of computer sciences and artificial intelligence to handle the mass of data. Biosphere 2 is a great 'university' for learning this discipline. And Biosphere 2 and the step into space are the greatest adventures around. Thus, 1 + 2 = 3."

C. ALLAN MORGAN

Roy Walford — born San Diego, California, USA, 1924.

C. ALLAN MORGAN

Bernd Zabel — born Munich, Germany, 1949.

Bernd Zabel

"Talent develops in quiet places, character in the full current of human life." Goethe's words were a strong influence for Bernd during his youth in Munich, at the foothills of the Bavarian Alps.

His childhood home was a haven for engineering talents, as his engineering father was Managing Director of the Siemens corporation. Bernd built his engineering talent quietly in his workshops, beginning with inventions in the family basement. He was also enthralled with the mysteries of nature and always had some biological experiment going as well.

He graduated with a Master's degree in electrical engineering from the Technical University in Munich. One semester during the revolutionary times of the late sixties, the entire school was closed by a strike. Bernd packed up his old Volkswagen bus and with several friends spent five months travelling and camping in Greece, Turkey, Iran, Iraq, Afghanistan and Pakistan. His courage and resourcefulness pulled them through many a scrape!

After studying philosophy for a year at the Ludwig Maximilian University in Munich, he spent four years getting a degree in education. Then time once more for travelling: from Tanzania to Egypt, to experience the life of the people and their cultures.

Back in Germany, he began teaching electrical engineering. It was a time when people were becoming aware that acid rain was decimating the legendary Black Forest and pollution was taking its toll on the once green and plentiful ecology of Bavaria.

Next year, he ventured to America and visited a German friend who was living at Santa Fe, New Mexico. There he made contact with Mark Nelson and used all of his vacation experimenting with solar energy and wind energy systems. He was hooked, and soon came again, to work with Mark over several years on aquaculture, drip-irrigation, water harvesting micro-catchments and other innovative agricultural techniques. Learning of Biosphere 2, he applied as a candidate.

He became manager responsible for the coordination and implementation of all construction on Biosphere 2, which had to be coordinated with the movement in of plants and animals from around the world. He was also largely responsible for the development of the fish-azolla-rice aquaculture subsystem, a satisfying and productive harmony of technics and biology.

In his spare time, he has become an accomplished painter and plans to continue this activity in Biosphere 2 over the next two years. Does he expect these two years to be a time of quiet development of talent, or of the rigors of character development? "Ideally, I am ready for anything, I hope!"

Biospherians, from left to right:
Standing, Tabor MacCallum,
Sally Silverstone, Linda Leigh,
Mark Van Thillo, Mark
Nelson. Seated, Jane Poynter,
Abigail Alling, Roy Walford.

At sunrise on September 26, 1991, the biospherians entered Biosphere 2 to begin their unique journey into the heart of life. One of the original team, Bernd Zabel, looked on from the outside. Not long before closure, it was found out that he had developed a medical problem that would preclude him from the first crew. It was with great regret that he agreed to withdraw from the team. Bernd continues to be active as a key member of the Mission Control Team, continuing his training for a later experiment and working with the waste recycle systems based on the Biosphere 2 design which will serve the Visitors Center. Mark Nelson, who replaced Bernd Zabel, hails from Brooklyn and his story is on the back of this page.

Each of the biospherians has a counterpart on the outside, so there is a dialogue on every area of responsibility. Comparisons are being made between Biosphere 2 ecosystems and their analogs in the Biospheric Research and Development Center greenhouses and labs. Through Mission Control, a team of colleagues and specialists keeps in close contact with the biospherians. Around Biosphere 2, a new complex is in operation, dedicated to acquainting the public with biospherics and designed to answer the questions of the hundreds of thousands drawn each year by the beauty and significance of this experiment.

Although separated from us by only a thin enclosure, and exchanging data, thoughts and emotions by computer, telephone and video, the biospherians are, truly, living in another world with a different atmosphere and different rhythms. Their goal: to make a major step in discovering the real forces which sustain the survival and evolution of life.

Mark Nelson, born Brooklyn, New York, 1947

A first generation American, Mark was born of Russian and Polish Jewish parents. He left New York for Dartmouth College and graduated with high honors in Philosophy, also taking pre-medical science courses. Back in New York he worked as social worker, court reporter and taxi driver.

His enquiring mind led him towards the arid southwest. There, he became interested in both ancient and modern ecological techniques. In Santa Fe he set up and managed an orchard, planting over two thousand fruit, shade and windbreak trees to reverse the local trends of desertification, improving the soil, conserving water and creating a veritable oasis. During this time he worked with Hopi Indian farmers and in the Negev desert in Israel.

Mark was co-founder in 1973 of the Institute of Ecotechnics, a non-salaried think-tank devoted to learning how to harmonize man's techniques with his ecological environment. The Institute consults to companies doing innovative projects in a number of diverse ecologies—from rainforest to coral reef.

In 1978, Mark's interest in arid zone ecology projects took him to the remote outback of Australia. He helped start a project in the overgrazed Kimberleys of Western Australia, a 5000 acre enterprise concerned with the regeneration of pasture land.

In 1984, he was one of the people that conceived of the Biosphere 2 project and the formation of Space Biospheres Ventures. Mark took charge of Space Applications, a theme which he has pursued ever since. From 1986 to 1991, he worked on the development of the Biosphere 2 agriculture system.

In 1987 he organized and chaired the first International Workshop on Closed Ecological Systems, held at the premises of the Royal Society in London. In 1989, he chaired the joint NASA/SBV workshop on Biological Life Support Systems. With John Allen and a few others, he was amongst the first Westerners to gain access to the Bios-3 project, which is the most advanced Soviet facility for research in closed ecological systems. This was in the Siberian city of Krasnoyarsk, where the second international Workshop on Closed Ecological Systems was held in 1989. The year before, Mark had become a founder-member of the Vernadsky Foundation.

Although active as conference organizer, writer and speaker, every year he has continued to work on the projects that he started in New Mexico and Australia. Now, this is on hold for two years. While in Tokyo as a keynote speaker at the first Japanese "Biospheres and Space" conference in May 1991, the phone call came offering him a place on the biospherian team.

He says, "It has always pained me that the history of many civilizations is one of ecological devastation. The prospect of humanity learning to become not desert-makers but oasis-creators represents for me not only an exciting challenge but our historical duty."

*The eight Biospherians together
on the beach in Biosphere 2.*

12
Futures

"The first thing that would impress a visitor from outer space would be the tremendous, inexplicable gap between potential and performance. It's amazing when you consider what the human organism could do in terms of its potential, and what it actually does. No species that isn't fundamentally flawed could be so stupid this consistently. Let us consider the human organism as an artifact. Comparative evolution will show us what is wrong with it and how far it has to go."

William S. Burroughs

Space-Faring

In the not so distant future, the adventurous scouts pushing outward on the edges of human horizons may have us a space-faring civilization. That is, if we can marshall the resources needed for such a giant thrust of human initiative. Some suspect that the time slot during which those resources will be available is fast narrowing. Before too long, the burdens of population explosion, agricultural stress on the environment, technological paralysis, mineral and fossil fuel depletion, and numerous other potentially debilitating conditions of an overburdened planet will prevent the expansion of the human species out into the Solar System. While the window of opportunity is still open, while money and other resources are still available, SBV pushes ahead to develop some of the science and biotechnics necessary for a permanent presence of our species off the mother planet. And perhaps the effort to master biospheric laws may teach us at the same time how to avoid the deterioration of the life support system of Earth.

NASA

Above: Earth seen from the far side of the moon.

Left: Jack McCauley of the USGS explaining Mars geology and possible sites for Mars colonies to SBV. McCauley contributed to NASA geological work on Mars, and trained astronauts for their Moon geological missions.

Opposite: The full moon beckons above the glimmering structure of Biosphere 2.

MARIE ALLEN

SBV is interested in becoming a supplier of life-support technologies and systems for space exploration. The bioregenerative technologies will make going into space more economic and permanent. Of particular concern are the possibilities for building a permanent lunar base, for exploring Mars, supplying orbital space stations with bioregenerative life support systems, and, eventually, having a long-term Martian base.

© 1991 SPACE BIOSPHERES VENTURES

Little by little the dream comes closer to reality. Almost within reach is the launching of Space Station *Freedom,* a goal reaffirmed and encouraged by President Bush in 1989 during the celebrations of the twentieth anniversary of the Apollo landing on the Moon. The original plan of the National Aeronautics and Space Administration (NASA), during the Reagan administration, was to launch the space station in 1992. SBV wanted to be ready with the results of their first two year run by that date. They intend to revise and streamline the technology based on results from the first two year experiment, keeping in touch with the space community on how component parts and small systems based on their work can tie-in with other work and future space missions. President Bush also announced during the Apollo festivities that the United States would have a permanent lunar base on the Moon early within the next century, another project in which SBV hopes to participate.

SBV is now preparing experiments in opaque conditions requiring artificial light for applications of biospheric knowledge in micro-gravity. This model is a possible use of the external fuel tank portion of the space shuttles which are currently jettisoned to burn up in the atmosphere when empty.

1992 has been designated *International Space Year (ISY),* and the focus of the American public will be directed to the future of the space program. "Mission to Planet Earth" is the theme selected by the ISY Committee. Both the National Commission on Space in 1985 and an in-house NASA report put together by astronaut Sally Ride recognized that the United States space program needs to understand biospherics and closed ecological systems if the U.S. is to have a future in space. The phrase "Mission to Planet Earth" was used by Ride in the NASA report and then became a catchy title for a new effort using space-based capabilities — satellites — to understand our biosphere. This ten-year long study from a new array of permanently orbiting space observatories circling the globe will undoubtedly be a major component of the space program of the 1990s with enormous impact on how to solve our current problems in managing Earth's pollution, ozone layer, agriculture, marine life, rainforests, etc. There are also international programs with Japan and Europe launching satellites as part of NASA's plan for the Earth Observation Satellite (EOS) program.

The other significance of the year 1992, that it marks the 500th anniversary of Columbus's discovery of the New World, is not lost on SBV. Biosphere 2 leads the way to new worlds that are an even greater stretch for the human species than crossing the Atlantic in 1492 was for the Europeans. Mark Nelson, SBV's Director of Space Applications, sees it through a visionary's eyes: "In a way, maybe it's one piece of our destiny. I contend we have an obligation to help the biosphere find another evolutionary niche, to move off this planet." He points out that in spite of the damage humans have done to the planet, only the species *Homo sapiens* has the technical ability to assist the rest of their comrades of the biosphere in lifting off the planet. He sees a kind of redemption for our past offenses if we could achieve this new phase of the biosphere's evolution. "If we could seed our solar system with a biosphere, then we've ensured a much more diverse area for life to do its wonderful work of evolution, creating free energy and organized complexity, beauty, intelligence. I think it's rather a grand thing, a great prospect that humanity should take a lot of pleasure in achieving. Even if there turns out to be no immediate payback, we need to master biospherics. Humanity is a pioneering species — we require challenges and dreams to engage our energies."

Biosphere 2, the only prototype to date for research and development of extraterrestrial space colonies, sits on the threshold of that step. President Bush's third announcement during last year's celebrations, along with his plan to have Space Station Freedom in orbit by 2000 and a permanent lunar base soon thereafter, was his administration's commitment to a manned expedition to Mars by 2020. His commitment is actually a revival of the old goal first envisioned by NASA back in the 1960s, the final purpose of the space shuttle, the space station and the lunar bases: to get to Mars. The President did not commit himself to a permanent settlement, however. It is the Soviet space program that takes as its motto: "We must grow our own apples on Mars." Aside from various space societies and individuals, SBV may be the only business organization that echoes and even expands that declaration.

The importance of permanent biospheres and settlements on Mars is that they will mark an expansion of life to an area sufficiently distanced from the mother planet and so rich in territory, resources and

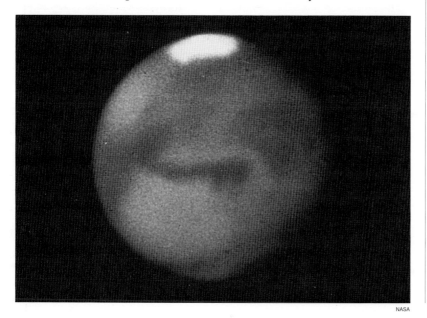

NASA

The gleaming north polar cap of Mars: its water and carbon dioxide are the raw materials of life.

149

Right: An SBV-Sarbid preliminary design for Mars Science Station and Hospitality base.

Right: An SBV-Sarbid preliminary design for Mars Science Station and Hospitality base.

Below right: The proposed floor plan for a Mars Science Station and Hospitality base.

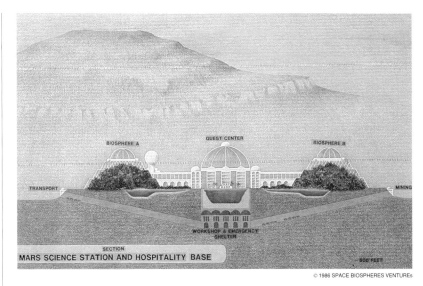

© 1986 SPACE BIOSPHERES VENTUREs

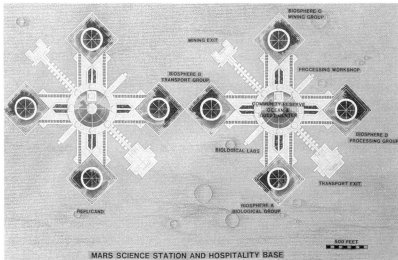

© 1986 SPACE BIOSPHERES VENTURES

opportunity, that a second center of permanent significance for life in our solar system will be established. A Moon base and the space station are seen as important positions for deep space launches and places to learn to live in space. But probably no decisive, irrevocable move will be made into the era of space biospheres without the creation of a Mars base. SBV has been working on designs for such a base.

One plan envisions a cooperating system of four biospheres connected by a central reserve ocean in its center. Each biosphere has all the biological and other additional systems needed to support from six to ten people. The hub area enclosing the reserve ocean would also be the center for the colony's community life and would contain a hospitality facility for visitors. Each of the four biospheres would have a particular function within the colony. A biological group would concentrate on ecology, agriculture, and medicine. A transport group would specialize in astronautics, machinery, and construction. A mining group would be responsible for geologic concerns, the extracting, concentrating, and refining of Martian resources, and in radioactivity. The processing group would handle chemical engineering, electronics, and manufacturing. All would play a part in the main objective: "to create and master the art of living on Mars."

Far Horizons

The dream of Biosphere 2 is perhaps as multifold as the people engaged in creating and inspiring it with life. A biosphere, by definition, contains multitudes. Just as the poet Walt Whitman sang in praise of diversity and grandeur, even when attended by paradox and contradiction, so too a generous future requires multiple horizons — to ensure not only tolerance of diversity as Jefferson wished, protection for diversity as Kennedy wished, but embracing the necessity of diversity in such worlds and futures as we continue to make and to dream.

This ability of humans to move off Earth and establish themselves in permanent communities would assure the evolutionary expansion of Earth's biospheric life as well as satisfy our insatiable instinct to explore. But the understanding of the development of ecosystems, enhanced through the study of biospherics, would bring more than a proliferation of spaceframe-enclosed structures dotting the landscape of Mars or, when the speed of space travel accelerates, the planets of other solar systems. It also would lead to the ability to change an entire planet to produce environments to support life, so that in some cases the settlers could open their doors whenever they felt like it or perhaps never even have to build a contained system on an alien planet to begin with.

"As radical an idea as this is, I would like to go even a step further," Frederick Turner, Founders Professor of Arts and Humanities at the University of Texas at Dallas, wrote in *Harper's Magazine* in August 1989, "Nature has had billions of years to try out different life-forms for their survival possibilities. Suppose evolution itself could be speeded up? Suppose researchers could create controlled conditions that would approximate those of Mars, and that, under such conditions, the Martian strains could, so to speak, generate themselves? The limits of metabolism make this impossible, the sceptic says. But the metabolic limits do not forbid software simulcra from *simulating* evolution. Contemporary computer biologists are already developing ways to express the genetic and somatic structures of simple Earthly organisms *as computer programs*. The next stage would be to place those programs together in a large cybernetic environment where they can compete and evolve for selective

Biosphere 2 silhouetted against the night sky.

151

fitness — and thus speed up the process of evolution electronically by many orders of magnitude." Inside the miracle world of computers, Turner suggests that the simulated conditions could gradually change from those of Earth's environment to those of Mars. Safe within the bounds of computer imaginings, the ensuing struggle for survival, one rapid generation after another, could be completely recorded and then used to plot the appropriate genetic engineering. And, voila! an instant ecosystem custom designed for Mars!

But Lovelock reminds us in his book, *The Ages of Gaia*, that even if such science fiction becomes reality, it's only the beginning. If Mars is to "become a self-sustaining system, it is necessary for the organisms and their environment to become as tightly coupled a system as they are on Earth. The acquisition of planetary control can come only from the growing together of life and its environment until they are a single and indivisible system."

Where might scientists test species from such supercharged evolution for a future place on Mars? In a closed ecosystem such as Biosphere 2, certainly, where the Martian conditions could be created. Already the SBV researchers and others are looking at pioneer candidates for Mars: the ginger plant, the banana, and the agave. It might take decades or centuries to condition Mars' temperature, atmosphere, and geology for the reception of plant life. But when it's ready, the first Martian plants, specially adapted to its conditions, may be waiting in a closed biosphere on Earth — or perhaps already on Mars!

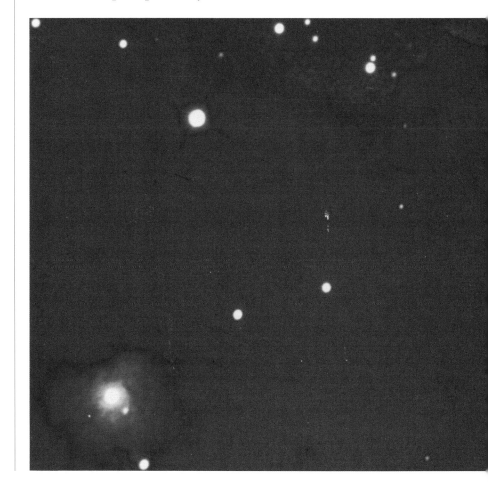

The Ultimate Human Experiment

As Biosphere 2 has begun its own life, a new confidence in and vision of the role of humanity in cosmic history is appearing. Man can no longer be seen as the measure of all things, or the conqueror of nature, or the species that became a parasite, but as a cooperative creative principal in the evolution of the cosmos with the extraordinary unlimited partner — the biosphere.

Our work had begun for real.
We had an apparatus that could work on a biospheric scale.
We could not only think and feel,
but sense and move upon the planet.
Everyone learned the arts of celestial navigation.
The seven continents and seven seas,
the Moon,
the Sun,
the planets and stars became the basic units of our vocabulary
which wind, water, rocks, life forms and cultures formed into words
which formed into texts
which formed into vision
which formed into Biosphere 2
which will transform into epics
changing the coordinates of reality.

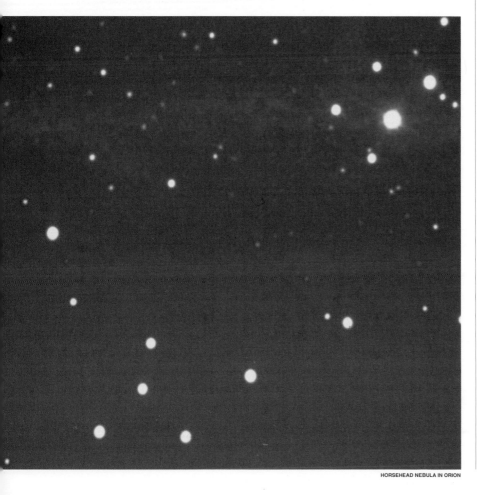

HORSEHEAD NEBULA IN ORION

Suggested Additional Reading

Allen, John & Nelson, Mark; *Space Biospheres*, Synergetic Press, 1990

Allen, John & Nelson, Mark; 'Ecology and Space' in *Biological Sciences in Space*, Vol.3 no.1, March 1989

Attenborough, David; *The Living Planet*, Little Brown, 1985

Augustine, Margret; 'Biosphere II - The Closed Ecology Project' in *The Human Quest in Space* , 1987

Biosphere, The, A Scientific American Book, W.H. Freeman & Co., 1970

Brand, Stewart (editor); *Space Colonies*, Penguin, 1977

Carson, Rachel; *Silent Spring*, Houghton Mifflin & Co., 1987

Commoner, Barry; *Making Peace with the Planet*, Pantheon Books, 1990

Darwin, Charles; *The Origin of Species*, New American Library, 1986

Dubos, Renee & Ward, Barbara; *Only One Earth: The Care and Maintenance of a Small Planet*, W.W. Norton & Co., 1972

Folsome, Clair; *The Origin of Life: A Warm Little Pond*, W.H. Freeman, 1979

Folsome, Clair & Hanson J.; 'The Emergence of Materially Closed System Ecology' in Polunin (ed) *Ecosystem Theory and Application*, John Wiley & Sons Ltd., 1986

Fuller, Buckminster; *Synergetics*, Macmillan & Co., 1983

Fuller, Buckminster; *Operating Manual for Spaceship Earth*, Simon & Schuster, 1989

Human Quest in Space, The, Proceedings of the twenty-fourth Goddard Memorial Symposium, AAS Science and Technology Series, 1987

Kamshilov, N.M.; *The Evolution of the Biosphere*, Mir Publishers, Moscow, 1976

Lapo, Andrey; *Traces of Bygone Biospheres*, MIR Publishers/Synergetic Press, 1987

Lorenz, Konrad; *Behind the Mirror*, Harcourt, Brace & Jovanovich, 1978

Lovelock, James; *The Ages of Gaia*, W.W. Norton & Co.,1988

'Managing Planet Earth', in *Scientific American*, September 1989

Maranto, Gina; 'Earth's First Visitors to Mars', in *Discover*, May 1987

McCourt, Richard; 'Creating Miniature Worlds', in *International Wildlife*, Jan-Feb 1988

Margulis, Lynn; *Early Life*, Jones and Bartlett, 1990

Margulis, Lynn & Sagan, Dorian; *Microcosmos: 4 Billion Years of Microbial Evolution from Our Bacterial Ancestors*, Summit Books, 1986

Margulis, Lynn & Sagan, Dorian; *Biospheres from Earth to Space*, Enslow Publishers, 1989

Margulis, Lynn & Schwartz, Karlene; *Five Kingdoms: An Illustrated Guide to the Phyla of Life on Earth*, W.H. Freeman & Co., 1982

Mumford, Lewis; *Technics and Civilization*, Harcourt, Brace & Jovanovich, 1963

Mumford, Lewis; *The City in History: Its Origins, Its Transformations and Its Prospects*, Harcourt, Brace & Jovanovich, 1968

Myers, Norman (general editor); *Gaia: An Atlas of Planet Management*, Anchor Press/Doubleday & Co., 1984

Nelson, Mark; 'The Biotechnology of Space Biospheres' in *Fundamentals of Space Biology*, edited by Asashiman & Malacinski, Springer-Verlag, 1990

Nelson, Mark & Soffen, Gerald (editors); *Biological Life Support Systems*, Synergetic Press/Space Biospheres Ventures, 1990

O'Neill, Gerard; *The High Frontier: Human Colonies in Space*, Bantam, 1978

Oberg, James E.; *New Earths*, New American Library, 1983

Odum, Eugene; *Fundamentals of Ecology*, Saunders College Publishing, 1971

Odum, Eugene; *Ecology: and Our Endangered Life-support System*, Sinauer Associates, 1989

Odum, H.T.; 'Limits of Remote Ecosystems Containing Man', in *American Biological Teacher* 25, 1963

Pearce, Peter; *Structure in Nature is a Strategy for Design*, MIT Press, 1990

Rambler, Mitchell; Margulis, Lynn & Fester, René (editors); *Global Ecology: Towards a Science of the Biosphere*, Academic Press, 1989

Pioneering the Space Frontier, National Commission on Space, Bantam, 1986

Remote Sensing of the Biosphere, National Academy Press, 1986

Sagan, Carl; *Cosmos*, Ballantine Books, 1985

Sagan, Dorian; *Biospheres*, McGraw-Hill, 1990

Sevastyonov, Ursul and Shkolenko; *The Universe and Civilization*, Progress Publishers, Moscow, 1981

Snyder, Tango Parrish (editor); *The Biosphere Catalogue*, Synergetic Press, 1985

Tsiolkovsky, K.S.; *Selected Writings*, Mir Publishers, Moscow, 1979

Vernadsky, Vladimir; *The Biosphere*, Synergetic Press 1986

Wilson, Edmund; *Biophilia*, Harvard University Press, 1984

Additional works cited in the text:

Asimov, Isaac; *Robot Novels*, Ballantine, 1988

Brand, Stewart; *Whole Earth Catalog*, 1967

Burroughs, William; *The Four Horsemen of the Apocalypse*, Expanded Media Editions, 1984

Clark, Arthur C.; *Songs of Distant Earth*, Ballantine, 1987

Goethe, G; *Faust*; translated by Caravan of Dreams Theater, in *Gilgamesh, Marouf the Cobbler, Faust Part 1*, Synergetic Press, 1984

Goff, Bruce; lecture notes taken by Phil Hawes

King, F.H; *Farmers of Forty Centuries: Permanent Agriculture in China, Korea and Japan*, 1911, reprinted Rodale Press

Kipling, Rudyard; *The Glory of the Garden*

Malone, Dumas; *Jefferson and the Ordeal of Liberty*, Little, Brown & Co., 1962

Sandburg, Carl; *The People, Yes*

Spretnak, Charlene; *Politics of Women's Spirituality*, Doubleday & Co., 1982

Suplee, Curt; 'Planet in a Bottle: the Making of Biosphere 2', in *The Washington Post*, January 21, 1990

Thompson, D'Arcy W.; *On Growth and Form*, edited by J.T. Bonner, Cambridge University Press, 1961

Todd, John; 'Architecture and Biology' in *Sustainable Communities*, Sierra Club Books, 1986

Turner, Frederick; 'Life on Mars' in *Harper's Magazine*, August 1989

Volovitch, Vitaly; *Experiment: Risk!*; (translated Beriozkina, Patty & Cogbill, Thomas), Progress Publishers, Moscow, 1986

Walford, Roy; *The 120-Year Diet*, Simon & Schuster, 1986

Walford, Roy; *Maximum Life-Span*, W. W. Norton & Co., 1983

Whitman, Walt; *Night on the Prairies*

Index

155